THE CONTINUATION OF ANCIENT MATHEMATICS

The Continuation of Ancient Mathematics

Wang Xiaotong's *Jigu suanjing*, Algebra and
Geometry in Seventh-Century China

Tina Su Lyn Lim
Donald B. Wagner

niasPRESS

The Continuation of Ancient Mathematics
Wang Xiaotong's Jigu suanjing, Algebra and Geometry in
Seventh-Century China
By Tina Su Lyn Lim and Donald B. Wagner

NIAS Reports # 51

First published in 2017 by NIAS Press
NIAS – Nordic Institute of Asian Studies
Øster Farimagsgade 5, 1353 Copenhagen K, Denmark
Tel: +45 3532 9503 • Fax: +45 3532 9549
E-mail: books@nias.ku.dk • Online: www.niaspress.dk

British Library Cataloguing in Publication Data
A CIP catalogue record for this book is available from the British Library

ISBN: 978-87-7694-217-5 (pbk)

Typesetting by Donald B. Wagner
Printed and bound in Great Britain
by Marston Book Services Limited, Oxfordshire

For Morten, Clara, and Annie

Contents

Information boxes

Figures

Preface

One of the major accomplishments of ancient Chinese mathematicians was the development of an algebra of polynomials. This began before the 1st century CE with algorithms for extracting square and cube roots. Later the arrangements on the counting board used in root extraction were recognized to be what we call polynomial equations, and algorithms were developed for the numerical solution of arbitrary polynomials. By the 14th century polynomial equations in up to four variables were being manipulated algebraically on the counting board to solve complex problems in astronomy and in the administration of labour for public works. After the 14th century the Chinese algebra of polynomials was largely forgotten; the reasons for this are obscure, but were no doubt related to the widespread substitution of the abacus for the counting board as well as administrative changes by the Ming dynasty (1368–1644). It may also be that 'pure mathematicians' had developed the method as far as it could reach without a paradigm shift, and gradually lost interest in it.

The book we are concerned with here, Wang Xiaotong's 'Continuation of ancient mathematics', shows us a stage in the development of the Chinese algebra of polynomials. Here in the 7th century the columns of numbers in the root-extraction procedures are recognized as equations which can be solved numerically, but these equations cannot yet be manipulated. Wang Xiaotong arrives at numerically solvable polynomials through a variety of *ad hoc* techniques, including geometric constructions and reasoning which may be classified as rhetorical algebra. Later, in the 18th century, Zhang Dunren would show that all of Wang Xiaotong's problems could be solved by straightforward manipulation of polynomials following the methods of the algebra of polynomials of the 14th century.

Wang Xiaotong lived in the late 6th and 7th centuries CE and served the Sui and Tang dynasties in posts related to calendrical calculations. He presented his book to the Imperial court at some time after 626, and in 656 it was made one of ten official 'canons' for mathematical education. The book contains 20 problems: one astronomical problem, then 13 on solid geometry, then six on right triangles. All but the first provide extensions of the methods in the mathematical classic *Jiuzhang suanshu* 九章算術 (perhaps 1st century CE): the solid-geometry methods in Chapter 5 and the right-triangle methods in Chapter 9, thus 'continuing' ancient mathematics. These require the extraction of a root of a cubic or (in two cases) quadratic equation.

The text includes comments in smaller characters which generally give explanations of the algorithms of the main text. These often amount to proofs of geometric propositions.

The core of this book is Tina's M.Sc. thesis in Mathematics at the University of Copenhagen, 2006, for which Don was one of the thesis advisers. We wish to extend special thanks to Karine Chemla for drawing our attention to Wang Xiaotong's interesting text, and to Jesper Lützen, Ivan Tafteberg, Jia-ming Ying 英家銘, and Jens Høyrup for useful advice and criticism along the way. The book includes material from previous articles (Lim and Wagner 2013a–b), and we are grateful to the editors and publishers for permission to use it here.

THE BOOK IS DIVIDED INTO three parts. Part I provides relevant background for an understanding of Wang Xiaotong's book. Part II discusses the mathematics of the book in detail, and Part III gives a full translation.

Tina Su Lyn Lim 林淑鈴
Donald B. Wagner 華道安
June 2017

Part I

Background

Part I provides relevant background material for an understanding of Wang Xiaotong's book. This includes its history, its function in practical applications and teaching, and the general mathematics of his time.

1.1. Wang Xiaotong and his book

Not much is known of Wang Xiaotong's life. We have the presentation of his book to the Throne, translated in Section 3.0 below, and we have the record of a controversy in which he was involved.[1] Beyond these there are brief mentions in other sources, usefully brought together by Li Yan (1963a: 82–83).

A modern sculptor's notion of Wang Xiaotong.

At the establishment of the Tang Dynasty in 618 it was necessary to produce a new system of calendrical calculations. The *Wuyin Calendar* 戊寅曆, compiled by a certain Fu Renjun 傅仁鈞, was proclaimed in that year, but two years later it began to show inaccuracies. An eclipse predicted for 620 did not occur. In 623 Wang Xiaotong was ordered to investigate, and he presented a revision in 626. Among other matters, he rejected the precession of the equinoxes and the variability of the speed of the sun along the ecliptic. This sparked a controversy which lasted until Li Chunfeng 李淳風 presented his *Linde Calendar* 麟德曆 in 665.[2] But Wang Xiaotong is not mentioned again in the record, and we need not follow the controversy further here.

On the basis of the sparse available sources, Li Di (1999) makes some educated guesses which are worth considering. Wang Xiaotong's report of 626 gives his title as Erudite of Calendrical Calculations (*suanli boshi* 算曆博士, a position not otherwise known).[3] The duties of Erudites were most often connected with education, so perhaps he taught calendrical science in the national universities (*guozi jian* 國子監) of the Sui and Tang Dynasties (Section 1.3 below).

Wang Xiaotong's memorial presenting his book (Section 3.0 below) states that his title is now Assistant to the Grand Astrologer (*taishi cheng* 太史丞). This was a post in the regular civil service, but with fairly low status, 28th rank (*cong qipin xia* 從七品下) out of 36.[4] His earlier position as Erudite would have ranked lower, perhaps 31st rank.[5] He states in the memorial that his report (dated 626) on the *Wuyin Calendar* was 'recent'; suppose therefore (on no other evidence) that the memorial was written in 628. At this time his 'hair is turning white', which was a conventional way of saying

---

1 *Jiu Tang shu* 1975, 32: 1153–1168, 79: 2710–2714; *Xin Tang shu* 1975, 25: 534–554. See also Section 3.0.4 below.

2 *Xin Tang shu* 1975, 26: 559.

3 *Xin Tang shu* 1975, 32: 1168.

4 *Jiu Tang shu* 1975, 42: 1798.

5 *Xin Tang shu* 1975, 47: 1216; des Rotours 1947: 212–213.

that he is getting on in years. If we suppose that he was in his 50th year, then he was born in 579, two years before the beginning of the Sui Dynasty (581–618); in his early career he would have served the Sui, and gone over to the Tang when this was opportune, perhaps a year or two before 618. Supposing (on no basis at all) that he died in his 60th year, we might write his dates as '(579? – 638?)'.

THE TITLE OF WANG XIAOTONG's book was originally *Jigu suanshu* 緝古算術, 'Continuation of ancient mathematics'.[1] Since 656, when it was made one of the ten official 'canons' (*jing* 經) for mathematical education (Section 1.3 below), it has been referred to as *Jigu suan***jing**.

The complicated history of the text will be treated in Section 1.6 below. In its present state it gives 20 mathematical problems, which Wang Xiaotong calls 'methods', *shu* 術. The first is astronomical, the next 13 are problems in solid geometry concerning the construction of earthworks and granaries, and the final six are abstract geometric problems on right triangles. The geometric problems can be seen as 'continuing' the mathematics of chapters 5 and 9 respectively of the classic *Jiuzhang suanshu* 九章算術 (Section 1.4 below).

The astronomical problem seems to be out of place here. Astronomy is not mentioned in Wang Xiaotong's presentation, and in its methods the problem is not related to the rest of the book; we wonder whether it really was part of the original book. For the sake of completeness we have translated it (Section 3.1 below), but have not otherwise dealt with its mathematics.

Problems 2–20 may be seen as exercises in solving a class of algebraic problems using either of two approaches, dissection of solid geometrical objects and 'reasoning about calculations'. These are discussed in Section 2.2 below. Each solution ends with a cubic equation (or in two cases a quadratic equation) to be solved numerically using the Chinese version of Horner's Method, discussed in Section 1.5 below.

Printed in smaller characters in the text are comments which generally give explanations of the algorithms of the main text. All extant editions state that the comments are by Wang Xiaotong himself, but we have noticed some differences in terminology between the comments and the main text,[2] and feel that the question of the

---

1 A more pedantically correct translation would be 'Computational methods which continue the ancient'.

2 E.g. in the commentary to problem 3, Section 3.3.3.4, fn. 1, p.148.

authorship of the comments should be left open. Perhaps most but not all of the comments are his.

THE *JIGU SUANJING* HAS NOT been much studied in modern times. There are two critical editions (Qian Baocong 1963; Guo Shuchun 1998), but these do not explicitly comment on the mathematical contents. Lin Yanquan (2001) translates the text into modern Chinese, expresses the calculations in modern notation, and gives derivations of some of the solutions. Deeper studies of individual parts of the text are by Shen Kangshen (1964), Qian Baocong (1966a), He Shaogeng (1989), Wang Rongbin (1990), Guo Shirong (1994), and Andrea Bréard (1999: 95–99, 333–336, 353–356). The comments in smaller characters in the text have hardly been studied at all by modern scholars: the most recent study we have found is that of Luo Tengfeng (1770–1841).[1]

---

1 Luo Tongfeng 1993. The only exceptions to this statement appear to be the highly speculative reconstructions of the fragmentary commentary on Problem 17 by He Shaogeng (1989) and Wang Rongbin (1990).

1.2. Public works planning in ancient China

Calculation was no doubt important throughout Chinese history in a wide variety of state and private activities, not least the keeping of accounts and the assessment of taxes. The ancient mathematical texts provide insight into a number of applications of mathematics, but in the broader historical sources we find only two: astronomy and the planning of public works. These are activities of officials at the highest level. In the available historical sources activities at lower levels in the official hierarchy, and those of private persons, are seldom visible.[1]

An interesting aspect of what we do know is that astronomy and public works went together: officials engaged in the one were often also engaged in the other (Wagner 2013). One example is Xu Shang 許商 in the first century BCE.[2] He was the author of two books (now lost) on calendrical calculations, but he is best known for some unfortunate advice which he gave on a public works project.

About 100 BCE the Yellow River had bifurcated, forming a tributary which was called the Tunshi River 屯氏河. In 39 BCE this river silted up; seven years later, the chief commandant of Qinghe Commandery 清河郡 (near modern Handan, Hebei), Feng Qun 馮逡, reported that the Yellow River was in danger of overflowing its dykes, and that dredging the Tunshi River would ease the pressure and reduce the danger of flooding.

> The memorial was passed to the Chancellor and the Imperial Counsellor. They responded that Xu Shang was an authority on the *Shang shu* 尚書,[3] that he was good at calculating, and that he could estimate the labour required. He was sent to inspect the situation, and reported that the Tunshi River was the cause of flooding, but that the local labour resources were

---

1 On the sources for the history of the Tang period and their limitations, see especially Twitchett 1979; note also 1992 and Balazs 1964. For a wider-ranging view of Tang scholarly activity, including history, see McMullen 1988.

2 On Xu Shang, see especially Loewe 2000: 622; Needham et al. 1971: 329–331; *Han shu* 1962, 9: 1686, 19b: 836, 841–842, 30: 1766.

3 Or *Shu jing* 書經, 'The book of documents', one of the earliest known Chinese books, with no relevance to the problem at hand. That Xu Shang was an authority on this book is mentioned to emphasize that he is a member of the cultured elite, not a mere technician.

insufficient, and dredging could be postponed. (*Han shu*
1962, 30: 1686; cf. Needham 1971: 329–331).

This was unfortunate advice, for three years later the Yellow River
overflowed its dykes and caused devastation, flooding 5000 square
kilometres to a depth up to 7 metres.

The author of the original memorial, Feng Qun, has a brief biography in the *Han shu* (1962, 79: 3305), seemingly only because his
family had a certain importance. He held various local administrative posts, rising to Governor of Longxi Commandery 隴西郡 (near
modern Lanzhou 蘭州, Gansu). It is probable that the initiative for
public works normally came, as in this case, from local officials at
lower levels of the bureaucracy. It was then the task of officials at
higher levels to organize a labour force for the project, drawing on
the population of more than one locality. This would often involve
calculating volumes of earth to be moved; then the number of man-
days required could be calculated and compared with the corvée labour available according to prevailing norms.

There are other anecdotes in the sources which tell of officials
using mathematics in the planning of public works, for example
a story of unwanted advice given by Zu Xuan 祖暅 (or Zu Geng
or Gengzhi 暅之, 5th century CE) and his subsequent imprisonment
(Wagner 2013). A Tang-period military manual describes the calculation of the labour requirements for the construction of a defence
wall.[1] Preserved fragments of the administrative codes of the Qin
(221–206 BCE) and Tang dynasties specify punishments for officials accidentally or intentionally giving incorrect estimates of labour requirements for construction projects.[2] But there is more detail
in the mathematical texts, to which we now turn.

1.2.1. Calculation of labour requirements
in the *Jiuzhang suanshu*

Determination of labour requirements would often involve, first,
calculation of the volume of a body of earth: a silted-up river, a canal
to be dug, a dam to be constructed of tamped earth. Then the number

---

1 *Taibai yinjing* 1854, 5: 1a–1b; on this book see Zhu Shida 2007: 7–16. Parallel text,
Tong dian 1988, 152: 3893.

2 *Shuihudi* 1978: 76–80; tr. Hulsewé 1985: 64; Li Jin 1985: 391; Liu Junwen 1983,
§240, 16: 312–313; 1996: 1208, 1210; tr. Johnson 1979–97, 2: 228–231; *Tiansheng
ling* 2006, §§15, 19, 26, 28: 422; Wang Maohua et al. 2012: 206.

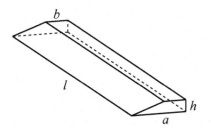

Figure 1. Dyke, *Jiuzhang suanshu*, Chapter 5, Problem 4.

of labourers needed could be calculated, given the current norms for corvée labour. The classic *Jiuzhang suanshu* 九章算術 ('Arithmetic in nine chapters', perhaps 1st century CE, see Section 1.4 below), has a chapter, 'Calculating labour quotas' (*Shang gong* 商功) with just such calculations. A simple example is Problem 4;[1] see Figure 1:

> A dyke has breadth 2 *zhang* 丈, upper breadth 8 *chi* 尺, height 4 *chi*, and length 12 *zhang* 7 *chi*. What is the volume?
>
> Answer: 7,112 [cubic] *chi*.

(In the Han period, 1 *zhang* = 10 *chi* ≈ 2.3 metres.)

The procedure for calculating this volume has already been given in Problem 1: It is equivalent to

$$V = \frac{(a+b)hl}{2} \tag{1.1}$$

In this case,

$$\frac{(20 \; chi \; + \; 8 \; chi) \; \times \; 4 \; chi \; \times \; 127 \; chi}{2} = 7,112 \; chi^3 \approx 85 \; \text{m}^3$$

Problem 4 continues:

> By the winter norm, one person's labour is 444 [cubic] *chi*. How many labourers are used?
>
> Answer: $16\,^2/_{11}$ persons.
>
> Method: Let the volume in [cubic] *chi* be the dividend and the norm in [cubic] *chi* be the divisor. Divide the dividend by the divisor; [the result] is the number of labourers.

The calculation is

$$\frac{7112 \; chi^3}{444 \; chi^3 \, / \, \text{labourer}} = 16\,^2/_{11} \; \text{labourers}$$

This labour calculation was simple, but Problem 21 complicates things.[2] It begins,

---

1   Qian Baocong 1963: 160; Chemla and Guo 2004: 388, 412–415; Guo Shuchun et al. 2013: 499–501.

2   Qian Baocong 1963: 171; Chemla and Guo 2004: 390, 440–443; Guo Shuchun et al. 2013: 593–599.

A pool has upper breadth 6 *zhang*, length 8 *zhang*, lower breadth 4 *zhang*, length 6 *zhang*, and depth 2 *zhang*. What is the volume?

Answer: 70,666 $^2/_3$ [cubic] *chi*.

The method for calculating this volume has been given under Problem 18. It is equivalent to (see Figure 2),

$$V = \frac{[(2l+k)a + (2k+l)b]d}{6} = 70,666\tfrac{2}{3}\ chi^3 \approx 860\ \text{m}^3$$

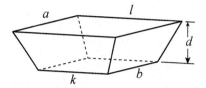

Figure 2. Pool, *Jiuzhang suanshu*, Chapter 5, Problem 21.

Problem 21 continues:

> The carrying of the earth is 70 *bu* 步 ['paces'], with 20 *bu* up and down wooden steps. Two [*bu*] on the steps correspond to five on a level path. For resting time, one is added for each 10. Time for loading and unloading is equivalent to 30 *bu*. One round trip is determined to be 140 *bu*. The capacity of a basket of earth is 1 *chi* 6 *cun* [i.e. 1.6 cubic *chi*]. The autumn norm for one person's labour is equivalent to walking 59 $^1/_2$ *li* 里.
>
> What is the volume [of earth] carried by each person, and how many labourers are used?
>
> Answer: One person carries 204 [cubic] *chi*, and the number of labourers is 346 $^{62}/_{153}$.

(In the Han period 1 *li* = 300 *bu* = 1800 *chi* ≈ 400 metres.)

The calculation given (not translated here) amounts to

$$\frac{10.6\ chi^3 \times 59\tfrac{1}{2}\ li\,/\,\text{person} \times 300\ bu\,/\,li}{\left[(70\ bu - 20\ bu) + \dfrac{5}{2} \times 20\ bu\right] \times 1.1} = 204\ chi^3\,/\,\text{person}$$

$$\approx 2.5\ \text{m}^3/\text{person}$$

$$\frac{70,660\tfrac{2}{3}\ chi^3}{204\ chi^3\,/\,\text{person}} = 346\,^{62}/_{153}\ \text{persons}$$

This is simple arithmetic, and the point of the problem seems to be to exercise the student's ability to organize a complicated calculation.

1.2.2. Wang Xiaotong on calculation of labour requirements

In the memorial introducing his book (Section 3.0.5 below), Wang Xiaotong criticizes current methods of calculating labour requirements:

I have humbly investigated the chapter 'Calculating labour quotas' of the *Jiuzhang* [*suanshu*]. Here there is a method for calculating the length on level ground [of a construction] assigned to [labourers owing] a quota of labour. But if [the construction in question] is wide above and narrow below, or high at the front and low at the back, the book lacks [a method] and does not consider it. As a result, modern persons, misunderstanding the deeper principles, treat crooked paths as if they were straight. This is 'putting a square handle into a round socket'; how could it be acceptable?

There are in fact no problems in the extant text of the *Jiuzhang suanshu* in which the *length* of a construction is calculated from the labour quota. It would have been easy enough, however, for a practical calculator in the Tang period to turn the formulas for volumes around, finding one dimension of a construction given its volume and the other dimensions. This would have been necessary, for example, if the dyke of Figure 1 above was to be constructed by several groups of workers, each group to construct a certain volume. Each group should then have been told the length of the part they were responsible for. From the formula for the volume of a dyke, Equation (1.1) above (p. 8), would be derived a calculation equivalent to

$$l = \frac{2V}{(a+b)h}$$

Things become more difficult if the dyke is 'wide above and narrow below, or high at the front and low at the back'. An example is Wang Xiaotong's Problem 3 (Sections 2.6.1.4 and 3.3 below), with a dyke which is not a simple prism but a more complicated object (Figure 33, p. 87), and is to be constructed by groups of workers from four counties. As can be seen there, the length of County A's contribution turns out to be a root of a cubic equation,

$$x^3 + 3{,}298\,^2/_{31}\,x^2 + 2{,}474{,}941\,^{29}/_{31}\,x$$
$$= 23{,}987{,}761{,}548\,^{12}/_{31}\,cun^3$$
$$x = 1{,}920\ cun \approx 576\ \text{m}$$

And this result appears to have been arrived at by using a volume dissection (Figure 34, p. 90).

SEVERAL OTHER PROBLEMS IN THE *Jigu suanjing* resemble this one in that they are, or can be construed to be, direct practical applications.

Others, however, are clearly not. For example, the first part of Problem 2 (Sections 2.5.1.1 and 3.2.1 below) gives the volume of a construction and seven differences between dimensions, and the dimensions are to be calculated: clearly not a practical problem. Even the 'practical' part of Problem 3, noted above, is not a real-world problem, for real dykes follow terrain rather than being strictly linear, and labour forces might also be expected to be more differentiated.

It has therefore been suggested to us that the *Jigu suanjing* is not a real textbook for practical calculation, but a *tour de force* in pure mathematics. This view comes from a misunderstanding of how textbooks work. No modern textbook of calculus for engineers, for example, gives real-world problems, for these are simply too complicated to serve as teaching examples. All of the geometric problems in the *Jigu suanjing* can be seen as simple examples of the methods needed in dealing with a particular class of practical calculations.

1.2.3. Digression: Calculation of labour requirements in the 14th century

A book of the Song and Yuan periods shows the kind of calculations which river conservancy officials in this time were expected to carry out. *Hefang tongyi* 河防通議, 'Comprehensive discussion of Yellow River conservancy', was originally written by Shen Li 沈立 in the 11th century and extensively edited by Shakeshi 沙克什 in the 14th century.[1] Discussions in the book include the history of the Yellow River, the administrative organs responsible for its control, engineering details on the construction of walls, dykes, and canals, and a whole chapter devoted to labour norms for various types of work. A final chapter on 'Calculation', *Suanfa* 算法, gives 27 calculation problems.

In the fifth of these problems we see a simple calculation concerning labour costs: 15,530 bundles of straw (much used in river conservation work) weighing 15 *jin* each are to be carried 90 *li*. The cost of carrying labour is 244 coins per 100 *li* and 100 *jin*. There are 1,000 coins in a string. The total expenditure is calculated as:

$$\frac{244 \text{ coins} \times 90 \text{ } li}{100 \text{ } li \times 100 \text{ } jin} \times 15 \text{ } jin / \text{ bundle} \times 15,350 \text{ bundles}$$

$$= 505 \text{ strings} + 629 \text{ coins}$$

---

1 For more on the complex history of this book see Guo Tao 1994; Guo Shuchun 1997; Wagner 2012b.

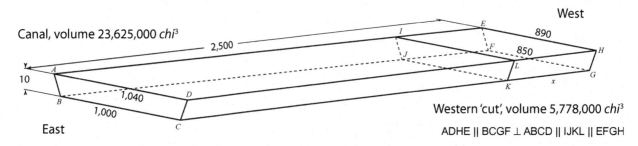

Figure 3. Canal, Problem 27 in *Hefang tongyi*, 14th century. Dimensions in *chi*. (At this time 1 *chi* ≈ 31 cm.)

More relevant to present concerns is the last problem, which resembles many of the problems in the *Jigu suanjing*. The length of a part of a canal constructed by one group of labourers is to be calculated. See Figure 3; the dimensions and the volume of the whole canal are given. From simple labour data the volume of the 'cut' at the western end can be calculated, and the length of the cut, *x*, is to be calculated. The length of the cut is a root of the equation

$$15x^2 + 94{,}500x = 2W = 11{,}556{,}000 \; chi^3$$

$$x = 120 \; bu$$

No doubt this result could have been arrived at by the same methods as in the *Jigu suanjing*, but here a more advanced method is used, the algebra of polynomials generally referred to as *tianyuan yi* 天元一, which will be discussed more fully in Section 2.10 below. The coefficients of polynomials were represented with rods on the counting board, in the same way as had long been done in the Chinese version of Horner's method, and were manipulated in much the same way as we manipulate equations algebraically; see Wagner 2012b for details as well as discussion of a philological problem which we have ignored here.

1.3. Mathematics education in Tang China

The *Jigu suanjing* was used as a textbook in the education of coming civil servants, and judging from Wang Xiaotong's preface it may have been written for precisely this purpose (Section 3.0.5 below). It is therefore necessary to look more closely at the organization of mathematics education in Tang China.[1]

On the teaching of elementary arithmetic our sources are sparse, but what there is suggests that many or most of the sons of the elite learned basic calculating skills in childhood. The more advanced mathematics which we meet in the extant texts seems before the Sui period (581–618) to have been transmitted in master–disciple traditions.[2] The Sui dynasty was probably the first to establish a mathematics specialization in its national university, the School for the Sons of the State (*Guozi xue* 國子學). At the beginning of the Tang dynasty (618) the mathematics specialization was abolished, but in 656 it was reestablished.[3]

A compendium on Tang administration, *Tang liudian* 唐六典, begun in 722 and completed in 739,[4] describes the organization of mathematics instruction in the national university as it was at that time (no doubt it changed in detail many times before and after this).[5] The staff included two Erudites (*boshi* 博士) in mathematics, 30 students, and two teaching assistants (*dian xue* 典學).[6] Erudites were members of the regular civil service, but their rank in the system was the lowest possible of 36 grades, *cong jiu xia* 從九下.

The university had six courses of study. Three (*xiucai* 秀才, *mingjing* 明經, and *jinshi* 進士) may be loosely classed as 'liberal arts', concerned with the classics, literature, and current affairs respectively, while three were more technical: law (*mingfa* 明法), calligraphy

---

1 This section leans heavily on the publications of Li Yan (1954), P. A. Herbert (1988), Taga (1985), David McMullen (1988: 35 ff); and – especially – Siu Man-Keung and Alexeï Volkov (1999). Note also Volkov 2012.

2 Li Yan 1954: 238–239; Siu and Volkov 1999: 87–88.

3 *Jiu Tang shu* 1975, 4: 76; *Xin Tang shu* 1975, 48: 1268.

4 des Rotours 1975; Wang Chao 1984; 1986.

5 *Tang liudian*, 21: 10a–11a. Parallel passages are found in *Jiu Tang shu* 1975, 44: 1892; *Xin Tang shu* 1975, 44: 1160–1161. The version in *Xin Tang shu* has been translated by des Rotours (1932: 139–140).

6 *Tang liu dian*, 21: 10a; cf. *Jiu Tang shu* 1975, 42: 1803; *Xin Tang shu* 1975, 48: 1268.

(*mingzi* 明字), and mathematics (*mingsuan* 明算).[1] It is interesting that these are the only practical subjects taught in the national university – where were astronomy and calendrical calculations learned? Where architecture, civil engineering, and a host of other subjects necessary for government? We have found no answer to any of these questions.

A description of the teaching of mathematics in the university is given in *Tang liu dian* and was copied into many other books of Tang history:[2]

> The Erudites in mathematics teach the sons of civil and military officials of the 8th rank and below and those of commoners who can support themselves. The textbooks[3] are divided into two courses of study. Fifteen students study [the following books]:
>
> *Jiuzhang* [*suanshu*] 九章[算術]
> *Haidao* [*suanjing*] 海島[算經]
> *Sunzi* [*suanjing*] 孫子[算經]
> *Wucao* [*suanjing*] 五曹[算經]
> *Zhang Qiujian* [*suanjing*] 張丘建[算經]
> *Xiahou Yang* [*suanjing*] 夏侯陽[算經]
> *Zhoubi* [*suanjing*] 周鞞[算經]
> [*Wujing suan*[*shu*] 五經算[術]][4]
>
> and fifteen students study these:
> *Zhui shu* 綴術
> **Jigu [suanjing] 緝古[算經]**
>
> These are studied together with [*Shushu*] *jiyi* [數術]記遺 and *Sandeng shu* 三等數.
>
> *Sunzi* and *Wucao* require together one year's study, *Jiuzhang* and *Haidao* together three years, *Zhang Qiujian* and *Xiahou Yang* each one year, *Zhoubi* and *Wujing suan* together one year, *Zhui shu* four years, *Jigu* [three][5] years.

---

1 Taga 1985: 27. We note that a diligent search has not produced the name of a single person holding the *mingsuan* degree. Their status was obviously not high, and thus they are invisible in the sources.

2 See fn. 5 on p. 13 above.

3 Or 'canons', *jing* 經.

4 Added from the parallel passage in *Xin Tang shu*.

5 Corrected by reference to *Xin Tang shu*.

The books listed here were edited by a group headed by Li Chunfeng 李淳風 (602–670);[1] thanks to their efforts, most are still extant today (the exceptions are *Xiahou Yang suanjing* and *Zhui shu*), while numerous other mathematical books which existed then are now lost.[2]

The eight books for the first programme appear to be oriented toward a cultural or historical understanding of mathematics.[3] Only the *Jiuzhang suanshu* is a major advanced mathematical text. *Wujing suanshu*, for example, gives some simple calculations related to passages in the Chinese classics, and *Zhoubi suanjing* is a Han text concerned with calculation of the dimensions of the universe according to a cosmology which by the Tang period had only historical interest.[4] If we knew more about the lost *Zhui shu* we might be able to decide whether the two programmes were in fact more culturally oriented on the one hand and more practical on the other.

1.3.1. *Zhui shu*

The *Zhui shu* mentioned here is a book by the famous mathematician and astronomer Zu Chongzhi 祖冲之 (429–500). Wang Xiaotong mentions it in his preface (Section 3.0.3 below), attributing it to Zu Chongzhi's son Zu Xuan 祖暅.[5] No doubt Zu Xuan edited or expanded his father's work. The book is long lost, and what it might have contained is a matter of controversy. Even the meaning of the title is uncertain.

The word *zhui* 綴 had in this time numerous meanings, including: sew, connect, join together; write, compose; writings; embellish, decorate. In addition, a 'technique of *zhui*', *zhuishu*, was later, in the Song and Yuan periods, used in astronomy.[6] It was probably

Artist's impression of Zu Chongzhi with his son Zu Xuan.

---

1 Exactly when this occurred is uncertain. Li Yan (1954: 249) quotes sources which respectively date the edition to the period 618–626, or 656, or 680.

2 See e.g. the bibliography in *Sui shu* (1973, 34: 1022–1026), completed in 636 CE, which unlike many similar bibliographies is explicitly a list of extant books.

3 For more on these books see e.g. Martzloff 1997: 123–141.

4 Cullen 1996.

5 Much later Shen Gua [or Kuo] 沈括 (1031–1095) also mentions a *Zhui shu* by Zu Xuan, giving the characters 祖亘 rather than 祖暅, presumably a simple scribal error (*Mengxi bitan*, Hu Daojing 1962, 18: 572; translation Hu Daojing et al. 2008: 531).

6 Shen Gua in *Mengxi bitan*, Hu Daojing 1962, 8: 334–335; translation Hu Daojing et al. 2008: 259–261; Qin Jiushao 秦九韶 (1202–1261) in *Shushu jiuzhang* 數書九章 (1842, 3: 21a–31b; 1936, 4: 79–91). The *zhuishu* problem in the latter book has been explicated by Qian Baocong (1966b: 92–95) among others. Libbrecht (1973: 470–472) gives a translation of Qian Baocong's explication, but typographical errors in the equations make the translation incomprehensible without reference to the original.

Zu Chongzhi, relief on the outer wall of Yandao Street Middle School in Chengdu, Sichuan. Photo DBW.

an advanced version of the technique of interpolation by finite differences, but whether this technique had anything to do with Zu Chongzhi's book, which seems to have been long lost by this time, is an unanswered question.

Li Chunfeng was the primary author of the 'Calendrical treatise' in the *Sui shu* 隨書 ('History of the Sui dynasty').[1] Here he wrote of Zu Chongzhi's calculation of π, 3.1415926 < π < 3.1415927:[2]

> Toward the end of the [Liu-]Song period [420–479], Zu Chongzhi, who held the position of Retainer Clerk in Nan Xuzhou 南徐州 [in modern Jiangsu province], calculated a better value. Taking the diameter of a circle, one *yi* 億 [10,000,000 *chi*], to be one *zhang* 丈 [10 *chi*] [i.e., dividing one *zhang* into ten million parts], the upper limit is 3 *zhang* 1 *chi* 4 *cun* 1 *fen* 5 *li* 9 *hao* 2 *miao* 7 *hu* and the lower limit is 3 *zhang* 1 *chi* 4 *cun* 1 *fen* 5 *li* 9 *hao* 2 *miao* 6 *hu*. The correct value lies between these limits. . . . He also established [the techniques] *kaichami* 開差冪 and *kaichali* 開差立 and used them in circle mensuration. He was concise and precise, among the greatest of mathematicians. His book was entitled *Zhui shu*. Among the educational officials there was none who could plumb its profundity, and for this reason it was abandoned and not maintained.

Qian Baocong (1963: 3) is surely correct in identifying *kaichami* and *kaichali* as the methods for numeric solution of quadratic and cubic equations ('Horner's method', see Section 1.5 below). What the last sentence in the quotation above implies is not at all clear: we suspect that it means that the seven-year course at the national university which used *Zhui shu* and *Jigu suanjing* was not a success and was discontinued. At any rate, when the 'Ten mathematical canons' were to be printed in 1084, the *Zhui shu* was no longer extant.

From Wang Xiaotong's memorial (Section 3.0.3 below) we know that *Zhui shu* contained something about 'entering a square city' (perhaps concerning right triangles) and something about two geometric solids, *chuting* and *fangting*. We have no further definite information on the contents of the book, but most scholars who have considered the question assume that it included three mathematical subjects which Zu Chongzhi and Zu Xuan are known to have written about: Zu Chongzhi's calculation of π (Volkov 1994),

---

1 Pu Qilong 1961, 12: 47, 48; *Jiu Tang shu* 1975, 79: 2718; Balazs 1953: 118–119.

2 *Sui shu* 1973, 16: 388.

Zu Xuan's derivation of the volume of a sphere (Wagner 1978b), and Horner's method.

We would add to this the interesting fact that Zu Chongzhi, in a memorial introducing his new *Daming* 大明 calendrical system, states that he has 'consulted both Chinese and foreign [books]' (*guanyao Hua Rong* 觀要華戎).[1] Undoubtedly 'foreign' refers here to translations of Indian mathematical and/or astronomical texts. The *Sui shu* bibliography lists three such books;[2] these are no longer extant, but one Chinese translation of an Indian astronomical text is extant, *Jiuzhi li* 九執曆 (Navagraha calendrical system).[3] It has been studied and translated into English by Yabuuti Kiyosi (1979).[4] We think it likely that Zu Chongzhi's knowledge of the precession of the equinoxes came from India. It is quite possible that he also learned some Greek mathematical methods, for example trigonometry, from Indian sources, and that he included these in his book. Such methods would have been very foreign to Chinese ways of approaching mathematical problems, and this could have contributed to difficulties which students and teachers experienced with the *Zhui shu*. But this speculation is based on very little concrete evidence.

Qian Baocong (1963: 4) suggests that *Zhui shu* was the title of Zu Chongzhi's edition of the *Jiuzhang suanshu*, and we see this as the most likely hypothesis.[5] Zu Chongzhi's biography in *Nan Qi shu* states that he 'created an extension to the *Jiuzhang* in several tens of chapters' (*Jiuzhang zao zhuishu shushi pian* 九章造綴述數十篇).[6] The phrase here translated 'extension' is *zhuishu* 綴述, i.e. almost the same as the book title, with 術, 'method, procedure' replaced by 述, 'relate, narrate'.

1.3.2. Jigu suanjing as a textbook

The *Jigu suanjing* clearly assumes in the reader a background in the techniques covered by the *Jiuzhang suanshu*, so the assumption that

---

1 *Nan Qi shu* 1972, 33: 903.

2 *Sui shu* 1973, 34: 1026.

3 The book is preserved in *Kaiyuan zhanjing* 開元占經, j. 104.

4 R. C. Gupta (2010) reviews many other examples of contact between Indian and Chinese mathematics.

5 Many years ago one of us wrote that the commentary on the *Jiuzhang suanshu* attributed to Liu Hui might in fact be a conflation of Liu Hui's and Zu Chongzhi's commentaries (Wagner 1978a: 211–212). We still consider this a possibility, but Qian Baocong's hypothesis is much more likely to be correct.

6 *Nan Qi shu* 1972, 52: 906.

the *Zhui shu* was an extensively annotated edition of that book would clarify the question of how and when this background was to be gained by students in the second of the two mathematics programmes. And the assumption that *Zhui shu* included 'Horner's method' for extracting the roots of polynomials would explain why *Jigu suanjing* does not include instructions for this. Under these assumptions the programme in mathematics which includes only *Zhui shu* and *Jigu suanjing* would seem to have been a concentrated course in advanced practical mathematical techniques.

For the degree examinations in mathematics the *Tang liu dian* describes two venues. Examination of students of the national university was conducted by the Ministry of Rites (*Lǐ bu* 禮部),[1] while candidates sent by provincial authorities were examined by the Ministry of Personnel (*Lì bu* 吏部).[2] The requirements of the two examinations appear to have been the same.

The *Xin Tang shu* has a description of the examinations in the Ministry of Rites which appears to draw on the *Tang liu dian* text, but also includes details from some other source or sources. Its description of the six courses of study in the national university finishes with the two programmes in mathematics in essentially the same words as the *Tang liu dian* text translated above (p. 14). The *Xin Tang shu* then gives a description of the examinations. The first part seems to apply to all students of the university, not only the students of mathematics.[3]

> In each ten-day period one holiday is granted. Before the holiday the Erudites hold an examination. With respect to *reading* [i.e. memorization], for each thousand characters one three-character quotation is given [and the student must complete the quotation]. With respect to *explication*, for each two thousand characters the general meaning of one item is asked; if two out of three items are answered correctly, the student passes; otherwise he is punished.
>
> At the end of the year a comprehensive oral examination is held on the year's work. For ten items the general meaning is required; those who answer eight correctly are graded superior, six middling, and five inferior. Students who three times are graded inferior and those who have been at the university

---

1 *Tang liu dian* 4: 3a–4b.

2 *Tang liu dian* 2: 26b–27a. On these two ministries see Hucker 1985: 306.

3 *Xin Tang shu* 44: 1161; cf. des Rotours 1932: 154–155.

for nine years (six years for students of law) without being presented for the degree examination are sent down. . . .

The text goes on with details of degree examinations for particular courses of study, ending with that in mathematics:

In mathematics, to pass [an item in the examination], the candidate must state the general meaning of the item in a dialogue, elucidate the numbers, construct an algorithm (*shu* 術), and explain in detail the principles behind the algorithm.

[For the first of the two programmes] the examination includes three items from the *Jiuzhang* [*suanshu*] and one from each of . . . [the other seven books in the programme]. Six passes out of the ten are required. On the [*Shushu*] *jiyi* and the *Sandeng shu*, ten [three-character] quotations are given [and the candidate must complete] nine to obtain the degree.

[For the second of the two programmes], in the examination on *Zhui shu* and *Jigu* [*suanjing*], in the dialogue on the general meaning, when the candidate 'elucidates the numbers, constructs an algorithm, and explains in detail the principles behind the algorithm', in cases where there is no comment [*zhu* 注], they are required to combine [? *he* 合] the numbers and construct an algorithm without erring in its principles. Only then do they pass [on the item]. [They are examined on] seven items from *Zhui shu* and three items from *Jigu* [*suanjing*], and must pass six out of the ten. On the [*Shushu*] *jiyi* and the *Sandeng shu*, of ten [three-character] quotations [the candidate must complete] nine in order to obtain the degree.

Siu and Volkov (1999: 96) are surely correct in concluding that degree candidates were given ten problems to be solved using the methods of the given text. 'Constructing an algorithm' would mean stating the steps in the calculation. This would not be a simple matter of repeating a memorized text – memorization was tested in a separate examination – but of producing something new along the same lines as the text. For example, Chapter 8, *Fangcheng* 方程, of the *Jiuzhang suanshu* gives problems which involve setting up a matrix, representing what we call a set of linear equations, and solving the equations by what amounts to Gaussian elimination.[1] An examination problem based on this chapter might have a different

---

1 Qian Baocong 1963: 221–240; Chemla and Guo 2004: 599–658; Guo Shuchun et al. 2013: 904–1033.

number of unknowns from any problem in the chapter. In both the *Jiuzhang suanshu* and the *Jigu suanjing* there are also some very complicated calculations concerning labour norms (e.g. Problem 3, Section 3.3.1 below), and examination questions could easily have been variations on these.

The obscure statement about 'cases in which there is no comment' in *Zhui shu* and *Jigu suanjing* might possibly be explained as follows. Problems 2–14 in the *Jigu suanjing*, on solid geometry, are solved in either of two ways. Most are solved by dissecting a solid into parts whose volumes are sums of products of known quantities and powers of the unknown quantity. No comment is normally attached to these problems. Others are solved by a kind of algebra, 'reasoning about calculations' (Section 2.2.3 below), and these are supplied with detailed comments. We hypothesize that this latter type was considered the most difficult, and students were not expected to do more with this type of problem than apply the method in the book directly, while in the former they were expected to be more creative in solving the problem.

It is likely that the authors of the *Xin Tang shu* have severely abridged their source here;[1] if that source were extant we would surely have a better chance of understanding the examination system.

---

1 See e.g. des Rotours 1932: 141, fn. 1.

1.4. Mathematical background

So much has been published in English on classical Chinese mathematics that it seems unnecessary to dwell at length on the general historical background of Wang Xiaotong's mathematics.[1] A few topics must be addressed here, however, in order to keep our book self-contained.

The most important mathematical text of pre-Tang China is the *Jiuzhang suanshu* 九章算術, 'Arithmetic in nine chapters', from about the 1st century CE, which we have already mentioned several times here. It is a compendium of calculation techniques in a wide variety of fields, including simple arithmetic, root extraction, plane and solid geometry, and linear algebra.[2] It appears to have been studied by all pre-modern Chinese mathematicians.

A commentary by Liu Hui 劉徽 in the 3rd century CE explains the techniques of the *Jiuzhang suanshu* in ways which amount to proofs of propositions.[3] This commentary became a model for later mathematicians, including Wang Xiaotong.

In the following we take up three topics from the *Jiuzhang suanshu* and Liu Hui's commentary which are directly relevant as background for our study of Wang Xiaotong's mathematics.

Liu Hui on a Chinese postage stamp.

1.4.1. Calculation

The units of measure seen in the *Jigu suanjing* are defined in Box 3 on p. 36. More complete lists of units are given e.g. by Chemla and Guo (2004: inside front cover).

Calculation was done with 'counting rods', *chou* 籌, arranged and manipulated on a 'counting board'. The representation of numbers with counting rods is explained in Box 1 (p. 22). The elementary operations of arithmetic: addition, subtraction, multiplication, and division, have been well described by Lam Lay Yong and Ang Tian Se (2004: 53–78). These are essentially done in the same way as the arithmetic we

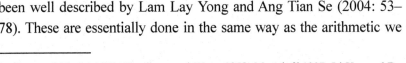

1 See e.g. Mikami 1913; Needham and Wang 1959; Martzloff 1997; Li Yan and Du Shiran 1987; and numerous pages on the World Wide Web.

2 Cullen 1993; Lam Lay Yong 1994. The best available translation in a Western language is Chemla and Guo 2004; others include Vogel 1968; Shen Kangshen et al. 1999; Guo Shuchun et al. 2013.

3 We speculate that Liu Hui's role here may have been in writing down the verbal commentaries of a long oral tradition. But evidence to confirm or negate this hypothesis is entirely lacking.

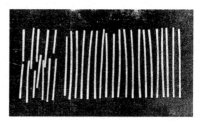

Bone counting rods from a Han tomb in Qianyang, Shaanxi. Reproduced from Lu Liancheng et al. 1976, pl. 1.

Box 1. Arithmetic with counting rods

Calculation with 'counting rods' (*chou* 籌) is mentioned in several historical sources as well as – of course – the classical mathematical treatises. Actual examples have been found in ancient graves ranging in date from the 3rd century BCE to the 3rd century CE.[1] They went out of use by about the 14th century CE, largely replaced by the abacus.[2]

Numbers were represented on the 'counting board' with two series of numerals:

1	2	3	4	5	6	7	8	9	0
—	=	≡	☰	☰	⊥	⊥	≛	≝	
I	II	III	IIII	IIIII	T	TT	TTT	TTTT	

Zero was usually represented by a blank position on the boards.

Numerals of the two series were used alternately, so that e.g. 2977 was represented by

$$= \mathrm{III} \perp \mathbb{T}$$

Negative numbers occur in intermediate results in the solution of systems of linear equations, and in this context Liu Hui states that *red* rods are used for positive numbers and *black* rods for negative numbers (Qian Baocong 1963: 225; Chemla and Guo 2004: 606–609, 625; Guo Shuchun et al. 2013: 931). But as Wang Qingjian (1993: 8) points out, this would be an extremely inconvenient way of indicating the sign of a number. It is likely that some simpler marker of sign was most often used on the counting board. Perhaps Liu Hui referred here not to the counting rods themselves but to colours in a now-lost illustration in his text.

---

1 See especially Du Shiran 1989; also Wang Qingjian 1993; Lu Liancheng et al. 1976; Li Shengwu and Guo Shuchun 1982; Zhang Pei 1988; 1992; 1996; Xiao Qi 1988. A tomb figure of a man identified as a 'calculator' (*jisuan yong* 計算俑) has been found in an Eastern Han tomb in Sichuan, but unfortunately the photograph in the excavation report is too poorly reproduced to reveal what the figure is doing (Jing Zhuyou 1993, fig. 11).

2 Li Yan 1963c; Lam Lay Yong 1987.

all learn in school; several manuscripts of the multiplication table are known from the Han and Tang periods (Li Yan 1963b; Libbrecht 1982). Division, however, has some interesting features.

In division, three rows on the counting board hold the quotient, dividend, and divisor. As each digit is added to the quotient, the product of its value and the divisor is subtracted, in one operation, from the divisor. This operation is called *chu* 除, 'eliminate', and division is referred to, *pars pro toto*, by the same word.

For example, in the division of 1297 by 16, successive states of the counting board would be:

quotient	0		8	〣	81	〣—
dividend	1297	|=〣⊥	17	|⊥	1	—
divisor	16	|⊥	16	|⊥	16	|⊥

The last shows that the result is 81 with a remainder of 1, or $81 \frac{1}{16}$, and numbers with fractions were always represented on the board as shown:

integer part

numerator

denominator

Or it was also possible to continuing dividing, forming a decimal fraction, 81.0625:

81.06	〣— ⊥	81.062	〣— ⊥ ||	81.0625	〣— ⊥||≣
40	≡	80	〣	0	
16	|⊥	16	|⊥	16	|⊥

Liu Hui uses decimal fractions in his calculation of π.[1] In giving his intermediate results he uses an extension of the system of linear measures shown in Box 3 (p. 36),

$$1 \text{ } chi \text{ 尺} = 10 \text{ } cun \text{ 寸} = 100 \text{ } fen \text{ 分} = 1{,}000 \text{ } li \text{ 釐}$$
$$= 10{,}000 \text{ } hao \text{ 毫} = 100{,}000 \text{ } miao \text{ 秒} = 1{,}000{,}000 \text{ } hu \text{ 忽}$$

By the thirteenth century these linear units developed into a general set of names for decimal places which could be used with any set of units, and there were names for decimal places up to 10^{-13} (Libbrecht 1973: 71–75). But for Liu Hui the notation stopped with *hu*, and this notation was inadequate to express his results. Therefore when, for example, in his calculation of π, he calculates the square root of 75

---

1 Qian Baocong 1963: 103–106; Chemla and Guo 2004: 145–149, 179–184; Guo Shuchun et al. 2013: 105–124.

square *cun* to six decimal places, yielding 8.660254 *cun*, he runs out of names for decimal places, and gives the last digit as a fraction:

8 *cun* 6 *fen* 6 *lí* 2 *miao* 5 $^2/_5$ *hu*

A FINAL POINT CONCERNING CALCULATION is the terminology of ratios in classical Chinese mathematics. The word *lü* 率 means one part of a ratio, so that for example the statement $\pi = {}^{22}/_7$ is expressed by saying that the *lü* of the circumference of a circle is 22 and the *lü* of the diameter is 7. We translate *lü* as 'proportion'. Chapter 2 of the *Jiuzhang suanshu* concerns calculations with proportions.[1]

1.4.2. Right triangles

In pre-modern Chinese mathematics the sides of a right triangle are referred to as *gou* 句 for the shorter leg, *gu* 股 for the longer leg, and *xian* 弦 for the hypotenuse (Figure 41, p. 102). We translate these terms as 'base', 'leg', and 'hypotenuse'.

Chapter 9 of the *Jiuzhang suanshu* gives 24 problems concerning right triangles. The first three present the Pythagorean theorem, while the rest are more complicated. An example is Problem 15:[2]

> If the base is 8 *bu* and the leg is 15 *bu*, what is the diameter of an inscribed circle?
>
> Answer: 6 *bu*.
>
> Method: With 8 *bu* for the base and 15 *bu* for the leg, calculate the hypotenuse. Add together the three quantities to make the divisor. Multiply the base by the leg and multiply by 2 to make the dividend. Divide to obtain the diameter in *bu*.

See Figure 4. This calculation is

$$d = \frac{2ab}{a+b+c}$$

Liu Hui proves this formula by a method of area dissection. He gives a diagram, now lost, which Qian Baocong (1958: 75) recon-

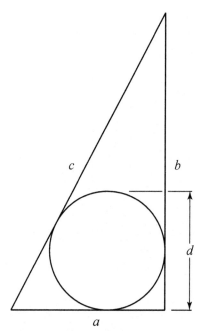

Figure 4. Circle inscribed in a right triangle, *Jiuzhang suanshu*, chapter 9, problem 15.

---

1 Qian Baocong 1963: 113–129; Chemla and Guo 2004: 199–219; Guo Shuchun et al. 2013: 172–255.

2 Qian Baocong 1963: 252–253; Chemla and Guo 2004: 726, 728–729; Guo Shuchun et al. 2013: 1110–1121.

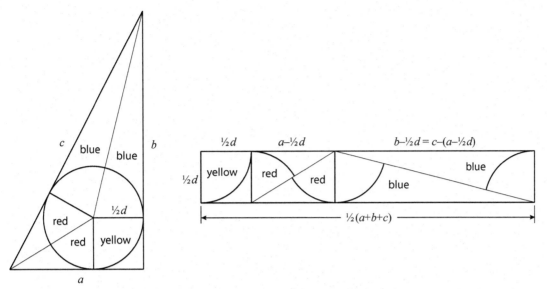

Figure 5. Derivation by dissection of the diameter of the inscribed circle.

structs as in Figure 5. The triangle is cut up and rearranged, showing that

$$\tfrac{1}{2}ab = \tfrac{1}{2}d\left(\tfrac{1}{2}d + [a - \tfrac{1}{2}d] + \tfrac{1}{2}\left[c - (a - \tfrac{1}{2}d) + (b - \tfrac{1}{2}d)\right]\right)$$

$$= \tfrac{1}{4}d(a + b + c)$$

and

$$d = \frac{2ab}{a + b + c}$$

1.4.3. Volumes of solids

Box 4 (pp. 38–39) shows the terms for geometric solids most often used in classical Chinese mathematics. The *Jiuzhang suanshu* gives formulas for their volumes, and Liu Hui gives derivations of these formulas.

Liu Hui derives the formulas for the *yangma* and *bienao* by a method which involves infinitesimal considerations (Wagner 1979). The other formulas build on this using volume dissections. An example is the formula for the volume of a *chumeng*,[1]

$$V = \frac{(2a + c)bh}{6}$$

---

1 Box 4, p. 38; Qian Baocong 1963: 169–170; Chemla and Guo 2004: 390, 437–439; Guo Shuchun et al. 2013: 575–581.

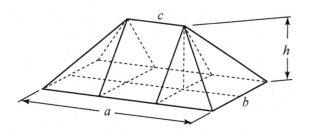

Figure 6. Dissection of a *chumeng*, *Jiuzhang shanshu*, chapter 5, problem 18.

Liu Hui divides the *chumeng* into four *yangma* and two *qiandu* as shown in Figure 6. He then constructs the box shown in Figure 7, which consists of twelve of the *qiandu* and eight boxes whose volumes are each three times the volume of the *yangma*. This corresponds to six times the number of *yangma* and *qiandu* in the original *chumeng*. The volume of the box is $(2a+c)bh$, and this is six times the volume of the *chumeng*. QED.

IN DEALING WITH CURVILINEAR SOLIDS Liu Hui implicitly uses a principle, never actually stated, which we may express in modern language as, 'if a solid with circular cross-section is inscribed in a solid with square cross-section, then the ratio of their volumes is equal to the ratio of the areas of the circle and square, $\pi : 4$'. Later Zu Xuan developed this further, to a principle (by Chinese historians generally called Zu Xuan's Axiom) which is essentially the same as Cavalieri's Theorem (Wagner 1978b; Lam Lay Yong and Shen Kangshen 1985). Liu Hui relates the *yuan baodao* (cylinder) to a box, the *yuanzhui* to a *fangzhui*, and the *yuanting* to a *fangting*.

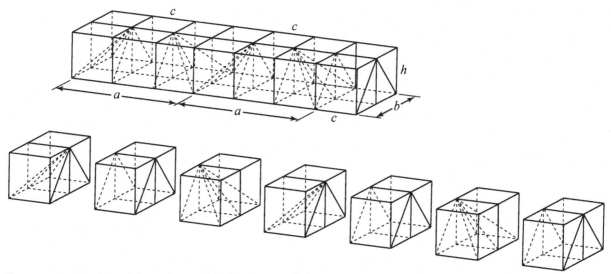

Figure 7. Derivation of the volume of a *chumeng*.

1.5. Extraction of the roots of cubic equations ('Horner's method')

Nearly all of Wang Xiaotong's problems involve finding a numeric root of a cubic equation. The equations are listed in Appendix 1 (p. 205); they all have the form

$$x^3 + ax^2 + bx = c \qquad (1.2)$$
$$a, b \geq 0, c > 0$$

Such an equation has exactly one positive real root (see Box 2, p. 29 below).

We have no information on exactly how Wang Xiaotong extracted the roots of these equations, but it was obviously a well-known procedure. Algorithms are described in detail in the *Jiuzhang suanshu* and the *Zhang Qiujian suanjing* 張邱建算經 (5th century CE) for the special case of extracting a cube root (the cubic equation $x^3 = c$), and extensions of the latter algorithm to general polynomials of all orders are given in several Chinese books from after Wang Xiaotong's time.[1]

This algorithm is equivalent to 'Horner's method' as it is sometimes taught in modern schools and colleges.[2] The first digit of the root is determined, the roots of the equation are reduced by the value of that digit (an operation which amounts to a change of variable), the next digit is determined, and this step is repeated until the desired precision is reached. The operation is carried out on paper in the Western version, in the Chinese version with calculating rods laid out on a table.

Reducing the roots of (1.2) by a value d gives a new equation,

$$y^3 + \left(a + 3d\right)y^2 + \left(b + 3d^2 + 2ad\right)y = c - bd - ad^2 - d^3 \qquad (1.3)$$

where $y = x+d$. The operation is continued with further digits until c is zero or very small (less than some predetermined ε). Classical Chinese mathematical texts call this operation *chu* 除, the same word used for the corresponding operation in long division (Section 1.4.1

---

1 Wang & Needham 1955; Lam Lay Yong 1970; 1977: 195–196, 251–285; 1986; Libbrecht 1973: 175–191; Chemla 1994; Martzloff 1997: 221–249; Shen Kangshen et al. 1999: 175–195, 204–226; Chemla & Guo 2004: 322–335, 363–379.

2 See e.g. Rees & Sparks 1967: 294–297 and numerous pages on the World Wide Web. Horner (1819) presented a procedure for approximating roots of any infinitely differentiable function, but modern descriptions of 'Horner's method' consider only the case of polynomial functions.

above), and here also the word is used *pars pro toto* for the whole operation of extracting roots.

The quantities on the counting board can be seen as variables in a computer program, and an efficient algorithm for obtaining (1.3) from (1.2) can be expressed in a computer programming language:

$$a = a+d \qquad \text{add } d \text{ to } a$$
$$b = b+d\times a \qquad \text{add } da \text{ to } b$$
$$c = c-d\times b \qquad \text{subtract } db \text{ from } c \qquad (1.4)$$
$$a = a+d \qquad \text{add } d \text{ to } a$$
$$b = b+d\times a \qquad \text{add } da \text{ to } b$$
$$a = a+d \qquad \text{add } d \text{ to } a$$

This algorithm includes only three multiplications, each by a single digit followed by zeroes.

Practical calculators were undoubtedly experienced in guessing the next digit in the root, and were seldom mistaken, but there is a method for finding the root without the benefit of experience. First, note that if d is too large, in the third line of (1.4), c will become negative. When the calculator discovers this, he can undo the calculations he has just done, reduce d by 1, and try again. Furthermore, it can be seen from (1.2) that the positive root must be less than $\sqrt[3]{c}$. This can easily be estimated, and a d chosen which is sure to be greater than or equal to the correct digit.

These considerations are brought together in the computer program, written in Basic, shown in Appendix 2. It can find the positive root of any equation of the form (1.2).[1]

A remaining problem is that several equations in the *Jigu suanjing* include fractions in the coefficients. How were these dealt with? In some, the denominators of the fractions are divisors of powers of ten, and these are unproblematic. For example, in Problem 17 (Section 3.17 below),

$$x^3 + 2\,{}^3/_4\,x^2 + 2\,{}^{21}/_{50}\,x = 812{,}591\,{}^{59}/_{125}$$

the fractions can be rewritten as decimals,

$$x^3 + 2.75\,x^2 + 2.42\,x = 812{,}591.472$$

---

1   There is more on the classical Chinese version of Horner's Method in Wagner 2017.

Box 2: Roots of a Wang Xiaotong cubic

We define a 'Wang Xiaotong cubic' to be a function of the form

$$f(x) = x^3 + ax^2 + bx - c \qquad \text{(a)}$$
$$a, b \geq 0, c > 0$$

and we shall prove that such a function has exactly one positive real root.

Because f in (a) is a cubic it must have at least one real root. This root cannot be zero.

Case 1. Suppose f has a negative root, $-r$. Then it can be factored as

$$f(x) = (x+r)(x^2 + \beta x - \gamma) \qquad \text{(b)}$$
$$\gamma > 0$$

The quadratic factor in (b) has two real roots, given by

$$\frac{-\beta \pm \sqrt{\beta^2 + 4\gamma}}{2}$$

These roots are both real, and since $\sqrt{\beta^2 + 4\gamma} > |\beta|$, one must be positive and one negative.

Case 2. If f does not have a negative root, then it must have at least one positive real root. If it has more than one positive real root, then it must have three, $r_1, r_2, r_3 > 0$, so that

$$f(x) = (x-r_1)(x-r_2)(x-r_3)$$
$$= (x-r_1)(x^2 - (r_2+r_3)x + r_2 r_3)$$
$$= x^3 - (r_1 + r_2 + r_3)x^2 + (r_1 r_2 + r_1 r_3 + r_2 r_3)x - r_1 r_2 r_3$$

whence

$$a = -(r_1 + r_2 + r_3) < 0$$

which contradicts (a). Thus f has exactly one positive root, which was to be proved.

---

and this gives no problems on the counting board, just as in long division. There are a few equations, however, which are not so easily handled, for example in Problem 11 (Section 3.11.3 below):

$$x^3 + 17\,^{37}/_{89}\, x^2 + 99\,^{14}/_{89}\, x = 6{,}429\,^{27}/_{89} \qquad \text{(1.5)}$$

It turns out that there is a fairly simple method of dealing with such equations on the counting board (see Section 3.3.3.3 below, p. 147). First, a common multiple D of the numerators is determined; in this case $D = 89$. Then the coefficients are converted to improper fractions in which the denominator of a is D, the denominator of b is D^2, and the denominator of c is D^3:

$$x^3 + \frac{1,550}{89}x^2 + \frac{784,193}{7,921}x = \frac{4,532,245,728}{704,969}$$

Then the denominators are discarded and the resulting equation is solved by the method given above,

$$x^3 + 1,550x^2 + 784,193x = 4,532,245,728$$

The root found, 1,157, is then divided by D to obtain the positive root of (1.5), $x = 13$. This procedure amounts to a change of variable, $y = Dx$.

Appendix 3 gives a Basic program to solve cubic equations in which fractions are treated as described here.

1.6. The history of the text of the *Jigu suanjing*[1]

The history of the text is discussed by the modern editors, Qian Baocong (1963: 490–491) and Guo Shuchun and Liu Dun (1998: 1: 21–22). As was mentioned in Section 1.3 above, the *Jigu suanjing* was one of ten books used in mathematics instruction which Li Chunfeng edited in the 7th century. These 'Ten mathematical classics' (*Suanjing shi shu* 算經十書) were printed for the first time in 1084. In the intervening centuries much had happened: two of the books, *Xiahou Yang suanjing*[2] and *Zhui shu* were no longer extant, and Li Chunfeng's commentary was missing from several. An exact copy (using the same number of characters per line and of lines per page) of that edition was printed in 1213 by Bao Hanzhi 鮑澣之.[3] By the 17th century the 1084 edition was no longer extant, and only part of the 1213 edition remained. The 1213 edition of *Jigu suanjing* existed at that time; it is no longer extant, but a hand copy prepared by Mao Yi 毛扆 (1640–?) and preserved in his Jigu Pavilion 汲古閣 survived into the 20th century. Qian Baocong appears to have seen the Mao Yi manuscript in the National Palace Museum, Beijing, perhaps in the 1930s. Its whereabouts are today unknown, but a facsimile reproduction was printed in 1931. This is the *Tianlu Linlang congshu* 天祿琳瑯叢書 edition, which is now available on the World Wide Web.[4]

There is good reason to believe that the extant text accurately reflects the text printed in 1084, but the relationship of this text to Wang Xiaotong's 7th-century text is a matter for speculation. It is marred by numerous banal scribal errors, and a reasonable guess is that the only copy available in 1084 was the result of generations of copying and recopying, perhaps by students who did not fully understand it.

The best Qing critical edition is that of Li Huang (1832);[5] others are by Dai Zhen (1777), Bao Tingbo (1780), Zhang Dunren (1803),

---

1 This section corrects several errors in our earlier articles (Lim and Wagner 2013a: 4; 2013b: 13–14).

2 The existing book with this name is not the original, but a later book substituted for it.

3 The extant fragments of the 1213 edition are reproduced in *Song ke suanjing liuzhong* 1981.

4 www.scribd.com/doc/78515037.

5 On Li Huang's edition see Huang Juncai 2009.

and the Korean mathematician Nam Pyŏng-Gil (1820–1869).[1] All of these are available on the Web. The standard modern edition has long been Qian Baocong's (1963, 2: 487–527; important corrections, 1966a), but that of Guo Shuchun and Liu Dun (1998) has much to recommend it. In the translation we have largely followed Qian Baocong's edition.

The many scribal errors must be corrected by reference to the mathematical context. Guo and Liu (1998, 1: 22) note that Li Huang introduced ca. 700 emendations to the text, and that Qian Baocong followed most of these but introduced 20 new emendations.

The end of the text (Problems 15–20, Sections 2.8, 3.15–3.20 below) gives special difficulties, for the last four double-pages of the 1213 copy available to Mao Yi were damaged, and numerous characters are missing.

ANOTHER MANUSCRIPT COPY OF THE *Jigu suanjing* is preserved in the National Central Library, Taibei. Professor Jia-ming Ying 英家銘 has kindly provided a copy;[2] we have compared this character by character with the *Tianlu Linlang* version, and found that the two are very nearly identical. There are two obvious scribal errors,[3] and one discrepancy which is more interesting. On the next-to-last page (p. 31b), where the 1213 edition was damaged and partly unreadable, the Taibei manuscript has one additional character, 須 (see fn. 1 in Section 3.19.1 below). This indicates that the manuscript was copied directly from the 1213 edition rather than from Mao Yi's copy, and that the copyist believed (rightly or wrongly) that he could discern one more character on the damaged page. Having two manucripts, independently copied and almost identical, we can conclude that the extant text is a very accurate copy of the 1213 version.

---

1 On Nam Pyŏng-Gil see Jia-ming Ying 2011; Horng Wann-sheng 2002; Zhang Fukai 2005.

2 A scan of the manuscript is at www.scribd.com/doc/285695980, and the entry in the library's catalogue is at rbook2.ncl.edu.tw/Search/SearchDetail?item=bd6f2a18ed814bac916f155604257d7cfDc0ODA00.

3 日 for 曰, p. 1b, and 十 for 一, p. 25a.

Part II

The Mathematics of the *Jigu suanjing*

This part analyses the mathematical techniques used by Wang Xiaotong. We consider the geometric problems, Problems 2–20, in an order which we find convenient, taking the simplest first and advancing through the more complicated problems. Sections 2.2, 2.9, and 2.10 take up Wang Xiaotong's methods more generally.

Problem 1, a simple pursuit problem, is translated in Section 3.1 below, but we wonder whether it really was part of Wang Xiaotong's original book, and find no reason to discuss it further.

It will be seen in the translations (Part III below) that the problems often include elaborate calculations concerning labour quotas. These were no doubt important exercises in organizing a large calculation, but the techniques used are simple arithmetic, and we do not treat them here.

2.1. Problems 7, 9, 13, and 14: Truncated square pyramids and cones

2.1.1. Problem 7: A square granary

This problem, translated in Section 3.7 below, has two parts. First the volume of 'a granary in the form of a *fangting*' (a truncated square pyramid, see Box 4, p. 39) is given, together with two relations between its dimensions; the dimensions are required. Second, a given amount of grain is removed, and the dimensions of the remaining grain are required. The first problem is solved using a volume dissection, the second using reasoning about calculations.

◆ *2.1.1.1. Calculation of the dimensions of the granary*

See Figure 8. The first part of the problem gives

$V = 187.2 \; hu \times 2.5 \; chi^3/hu = 468 \; chi^3$

$a - b = 6 \; chi$

$h - b = 9 \; chi$

(For these units of measure see Box 3, p. 36.)

In the calculation, one intermediate quantity is given a name, 'Area for the Corner *Yangma*'. We denote this quantity by K_1:

$$K_1 = \frac{(a-b)^2}{3} = 12 \; chi^2$$

This does not correspond to any 'area' in the geometric situation; it is merely an intermediate result in the calculation. Further calculations lead to the cubic equation,

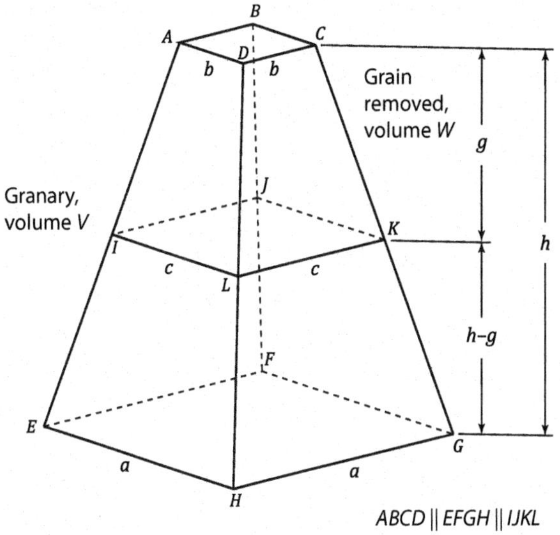

Figure 8. Square granary, Problem 7.

$$b^3 + \left[(a-b)+(h-b)\right]b^2$$
$$+ \left[(a-b)(h-b)+K_1\right]b = V - K_1(h-b) \tag{2.1}$$

$b^3 + 15b^2 + 66b = 360 \; chi^3$

This has one real root, $b = 3 \; chi$, and the remaining dimensions are

$$a = b + (a-b) = 9 \; chi$$
$$h = b + (h-b) = 12 \; chi$$

In this case the text gives no reasoning, only the calculation. However, the name given to the quantity which we here denote K_1, 'Area for the Corner *Yangma*', indicates that (2.1) was derived using a volume dissection like that shown in Figure 9. The 'Corner *Yangma*' are the four *yangma* labelled I.

The figure is divided into parts whose volumes are sums of products of known quantities and powers of b:

$$V_{\mathrm{I}} = \frac{(a-b)^2}{12} h = \frac{K_1}{4}(h-b) + \frac{K_1}{4}b \qquad (2.2)$$

$$V_{\mathrm{II}} = \frac{a-b}{4} hb = \frac{a-b}{4}(h-b)b + \frac{a-b}{4}b^2 \qquad (2.3)$$

$$V_{\mathrm{III}} = hb^2 = (h-b)b^2 + b^3 \qquad (2.4)$$

$$V = 4V_{\mathrm{I}} + 4V_{\mathrm{II}} + V_{\mathrm{III}}$$

Combining and collecting terms,

$$V = b^3 + \left[(a-b)+(h-b)\right]b^2 \\ + \left[(a-b)(h-b)+K_1\right]b + K_1(h-b)$$

which is equivalent to (2.1).

◆ *2.1.1.2. Calculation of the dimensions of the grain removed*

In the second part of the problem the volume of the grain removed, W, is given,

$$W = 50.4 \; hu \times 2.5 \; chi^3/hu = 126 \; chi^3$$

(see Figure 8) and the dimensions g and c are required.

In the calculation an intermediate quantity is given a name, 'Small Height', which we denote by K_2:

$$K_2 = \frac{hb}{a-b} = 6 \; chi \qquad (2.5)$$

Again this quantity corresponds to nothing in the geometric situation. The calculations lead to the cubic equation

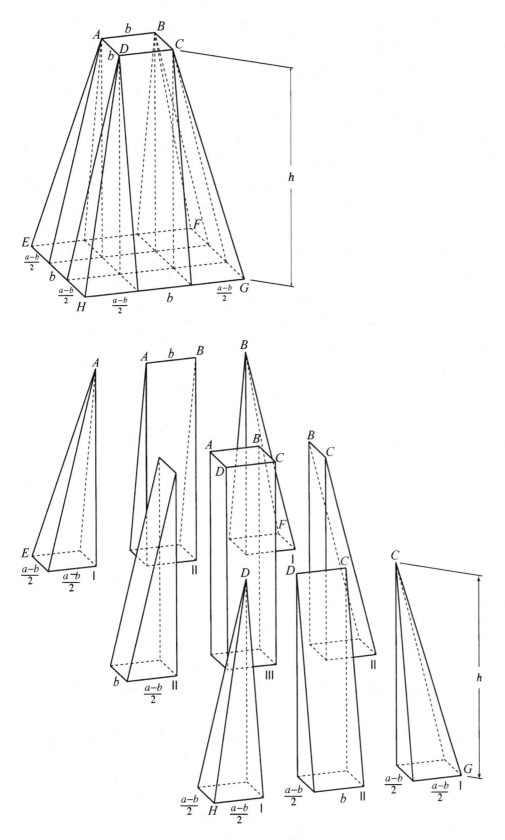

Figure 9. Dissection of the square granary, Problem 7.

Box 4: The standard geometric solids in classical Chinese mathematics

These are the basic geometric figures treated in Chapter 5 of the *Jiuzhang suanshu* (Qian Baocong 1963: 159–178; Chemla & Guo 2004: 388–390, 411–457; Guo Shuchun et al. 2013: 491–633).

Box: rectangular parallelepiped. There is no specific term for this, but once in *Jigu suanjing* (Section 3.17.1 below) a box is referred to as a *fang* 方 – a word with a wide variety of meanings, including 'square' and 'rectangle'.

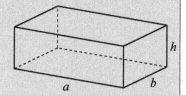

Volume = abh

Qiandu 塹堵: right prism with right-triangular base.

Volume = $\dfrac{abh}{2}$

Yangma 陽馬: pyramid with rectangular base and with one lateral edge perpendicular to the base.

Volume = $\dfrac{abh}{3}$

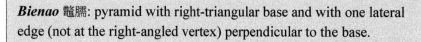

Bienao 鱉臑: pyramid with right-triangular base and with one lateral edge (not at the right-angled vertex) perpendicular to the base.

Volume = $\dfrac{abh}{6}$

Fangzhui 方錐: right pyramid with square base.

Volume = $\dfrac{a^2h}{3}$

Chumeng 芻甍: right wedge with rectangular base.

Volume = $\dfrac{(2a+c)bh}{6}$

Chutong 芻童: truncated right wedge with rectangular base.

Volume = $\dfrac{\left[(2a+c)b+(2c+a)d\right]h}{6}$

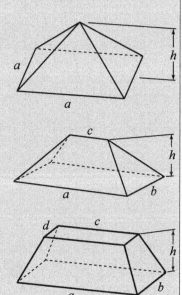

Fangting 方亭: truncated right pyramid with square base.

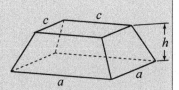

$$\text{Volume} = \frac{\left(ac + a^2 + c^2\right)h}{3}$$

Yanchu 羨除: wedge with trapezoidal base and with one lateral face perpendicular to the base.

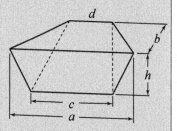

$$\text{Volume} = \frac{(a + c + d)bh}{6}$$

Curvilinear solids: In the *Jiuzhang suanshu* the value $\pi \approx 3$ is used. In the *Jigu suanjing* the value $\pi \approx 3$ is used once (Section 2.1.2 below), otherwise $\pi \approx {}^{22}/_7$ (Sections 2.1.1 and 2.1.3).

Yuan baodao 圓堢壔: cylinder.

$$\text{Volume} = \frac{\pi d^2 h}{4}$$

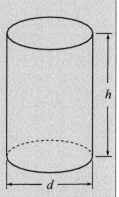

Yuanzhui 圓錐: right cone.

$$\text{Volume} = \frac{\pi d^2 h}{12}$$

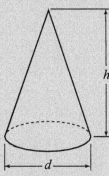

Yuanting 圓亭: truncated right cone.

$$\text{Volume} = \frac{\pi\left(cd + c^2 + d^2\right)h}{12}$$

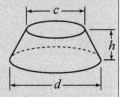

$$g^3 + 3K_2 g^2 + 3K_2{}^2 g = \frac{3Wh^2}{(a-b)^2} \tag{2.6}$$

$$g^3 + 18g^2 + 108g = 1512 \; chi^3$$

This has one real root, $g = 6 \; chi$. Then c is calculated, implicitly considering similar triangles:

$$c = \frac{g(a-b)}{h} + b = 6 \; chi \tag{2.7}$$

◆ *2.1.1.3. Derivation of the dimensions of the grain removed*

A derivation of (2.6) by volume dissection does not seem possible, and two comments in the text in smaller characters give a derivation by reasoning about calculations.

The main text calculates first the constant term of the cubic equation (2.6), then the coefficients of the linear and quadratic terms. Following the calculation of the constant term comes a comment in smaller characters (Section 3.7.3.2.2). This defines an additional named quantity, the Large Height, denoted here K_3, as 'the sum of the height of what was removed and the Small Height',

$$K_3 = g + K_2 \tag{2.8}$$

Note here that the Large Height is not known at this point in the text: g is the unknown quantity to be derived.[1]

The comment then states that the constant term in the equation (the *shi* 實) is the sum of 'the products of the Large and Small Heights each by itself and by each other and multiplied by the height of what was removed', that is,

$$shi = \frac{3Wh^2}{(a-b)^2} = gK_2{}^2 + gK_3{}^2 + gK_2 K_3 \tag{2.9}$$

A derivation of this statement will come in the second comment.

The main text continues with the rest of the calculation, and then another comment in smaller characters (Section 3.7.3.2.4) continues the derivation. This second comment starts by stating the formula for the volume of a *fangting*, given centuries earlier in the *Jiuzhang suanshu* (See Box 4, p. 39):

---

1 See fn. 1 in Section 2.5.1.2, p. 72 below.

> Here the original method is: [The sides of] the upper and
> lower squares [b, c] are multiplied by each other, and each is
> multiplied by itself. The sum is multiplied by the height [g]
> and divided by 3.

This corresponds to

$$W = \frac{\left(b^2 + c^2 + bc\right)g}{3} \qquad (2.10)$$

The derivation then manipulates this calculation.

> To return to the starting point, this is tripled, then multiplied
> by the area of [the square on] the height [h] and divided by
> the area of [the square on] the difference [a–b].

The result of this manipulation, not stated explicitly, is

$$
\begin{aligned}
shi &= \frac{3Wh^2}{(a-b)^2} = \frac{\left(b^2 + c^2 + bc\right)gh^2}{(a-b)^2} \\
&= \frac{gb^2h^2}{(a-b)^2} + \frac{gc^2h^2}{(a-b)^2} + \frac{gbch^2}{(a-b)^2}
\end{aligned}
\qquad (2.11)
$$

The comment then states (2.9) again,

$$shi = \frac{3Wh^2}{(a-b)^2} = gK_2^{\,2} + gK_3^{\,2} + gK_2K_3$$

and continues:

> Why is this? Multiplying the height [h] by [the side of] the
> lower square [c] and dividing by the difference between [the
> sides of] the squares [a, b] gives the Large Height [K_3].

This amounts to

$$K_3 = \frac{ch}{a-b} \qquad (2.12)$$

which is true because, considering similar triangles,

$$K_3 = g + K_2 = \frac{c-b}{a-b}h + \frac{b}{a-b}h$$

Then the definition of the Small Height is repeated:

$$K_2 = \frac{hb}{a-b} \tag{2.13}$$

and the comment continues with statements equivalent to

$$\frac{c^2h^2}{(a-b)^2} = \left(\frac{ch}{a-b}\right)^2 = K_3{}^2 \tag{2.14}$$

$$\frac{b^2h^2}{(a-b)^2} = \left(\frac{bh}{a-b}\right)^2 = K_2{}^2 \tag{2.15}$$

These follow from (2.12) and (2.13). Combining (2.11), (2.14), and (2.15) gives (2.9).

Then follows the proof of the cubic equation (2.6) using an area dissection, Figure 10. It is possible but not certain that the original book included a diagram like this.

> The product of the Large Height [K_3] by itself is a large square [$ABCD$]. Inside the large square are: one area [$DFGJ$] which is the product of the height of what was removed [g] by itself; one area [$BHGE$] at the corner which is the product of the Small Height [K_2] by itself; and two areas [$AEGF$ and $CJGH$] at the two sides which are [each] the product of the height of what was removed [g] by the Small Height [K_2].

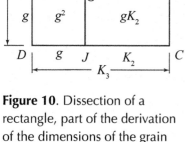

Figure 10. Dissection of a rectangle, part of the derivation of the dimensions of the grain removed.

This shows that

$$K_3{}^2 = (K_2+g)^2 = g^2 + K_2{}^2 + 2gK_2 \tag{2.16}$$

The comment continues:

> Further, the product of the Large Height [K_3] by the Small Height [K_2] is the middle rectangle [$BCJE$].[1] Inside the middle rectangle are one area [$CJGH$] which is the product of the Small Height [K_2] by the height of what was removed [g] and the area of the small square [$BHGE$] which is the product of the Small Height [K_2] by itself.

This shows that

$$K_2K_3 = K_2(K_2+g) = gK_2 + K_2{}^2 \tag{2.17}$$

---

1 Presumably *BCJE* in Figure 10 is called the 'middle rectangle' (*zhong fang* 中方) in comparison with the 'large square' (*da fang* 大方) *ABCD* and the 'small square' (*xiao fang* 小方) *FGJD*.

Combining (2.9), (2.16), and (2.17),

> Then the three areas given by the product of the Large and Small Heights [K_3, K_2] and their products with themselves are each multiplied by the height of what was removed [g] to make volumes. Therefore three times the area of [the square on] the Small [Height] [K_2] is the *fangfa* [the coefficient of the linear term] and three times the Small Height [K_2] is the *lianfa* [the coefficient of the quadratic term].

This amounts to (2.6),

$$g^3 + 3K_2 g^2 + 3K_2{}^2 g = \frac{3Wh^2}{(a-b)^2}$$

which was to be proved.

Granary, volume V

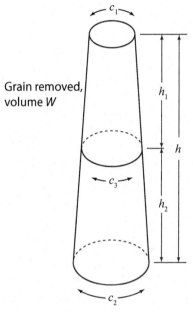

Figure 11. Circular granary, Problem 9.

2.1.2. Problem 9: A circular granary

In Problem 9 the volume of a truncated cone and two relations between dimensions are given, and the dimensions are required. Then a volume of grain is removed, and the dimensions of the empty space left are required. The problem is solved by converting it to one equivalent to Problem 7.

See Figure 11. The given quantities are

$V = 1764$ *chi³*

$c_2 - c_1 = 12$ *chi*

$h - c_1 = 18$ *chi*

$W = 666$ *chi³*

and 'the proportions to be used are diameter 1 circumference 3', i.e., the approximation $\pi \approx 3$ is to be used.

The calculation begins,

> Multiply [the volume of the granary] by 36 and divide by 3 to obtain the volume of a *fangting*.

See Figure 12. This is

$$F = \frac{36V}{3} \tag{2.18}$$

Fangting, volume F

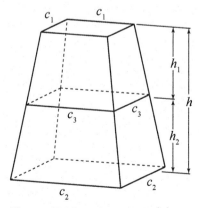

Figure 12. *Fangting*, part of the derivation of Problem 9.

The volume of this *fangting* is, using a version of Cavalieri's Theorem (Section 1.4.3 above),

$$F = \frac{c_1^2}{\left(\dfrac{c_1^2}{4\pi}\right)} V = 4\pi V \approx 12V$$

but the fact that the calculation multiplies by 36 and divides by 3 indicates that the author derived (2.18) by some more roundabout reasoning.

From here on the entire calculation is equivalent to that in Problem 7 (Section 2.1.1 above). The first equation whose root is to be extracted, corresponding to (2.1) above (p. 35), is

$$c_1^3 + \left[(c_2 - c_1) + (h - c_1)\right]c_1^2 + \left[(c_2 - c_1)(h - c_1) + K_1\right]c_1$$
$$= F - K_1(h - c_1)$$

where

$$K_1 = \frac{(c_2 - c_1)^2}{3} = 48 \; chi^2$$

Numerically,

$$c_1^3 + 30c_1^2 + 264c_1 = 20,304 \; chi^3$$

This has one real root, $c_1 = 18 \; chi$, and the other dimensions of the granary are

$$c_2 = c_1 + (c_2 - c_1) = 30 \; chi$$
$$h = c_1 + (h - c_1) = 36 \; chi$$

The second equation, corresponding to (2.6) above, is

$$h_1^3 + 3K_2^2 h_1^2 + 3K_2^2 h_1 = \frac{36Wh^2}{(c_2 - c_1)^2}$$

where

$$K_2 = \frac{hc_1}{c_2 - c_1} = 54 \; chi$$

Numerically,

$$h_1^3 + 162 \, h_1^2 + 8,748 \, h_1 = 215,784 \; chi^3$$

This has one real root, $h_1 = 18 \; chi$.

A COMMENT IN SMALLER CHARACTERS (Section 3.9.5) states that the *fangting* 'is not different from the previous rectangular storage pit', i.e. the object in Problem 6 (Section 2.5.2 below, Figure 28, p. 77), not in Problem 7. However, the object in Problem 6 is rectangular rather than square, so the calculation there is more complicated than necessary for the problem at hand.

2.1.3. Problems 13 and 14: Storage pits, square and circular

◆ 2.1.3.1. Problem 13: Two storage pits, one square and one circular

Problem 13 follows neatly on from the first part of Problems 7 and 9. We are given the total volume of two objects and two relations between their dimensions, and the dimensions are required.

See Figure 13; the given quantities are

$$V = 12{,}862.5 \ chi^3$$
$$a{-}h = 7 \ chi$$
$$h{-}b = 14 \ chi$$

and the approximation $\pi \approx {}^{22}/_7$ is to be used.

◆ 2.1.3.2. The calculation

The two figures are, in Chinese terminology (Box 4, pp. 38–39), a *fangting* (truncated square pyramid) and a *yuanting* (truncated

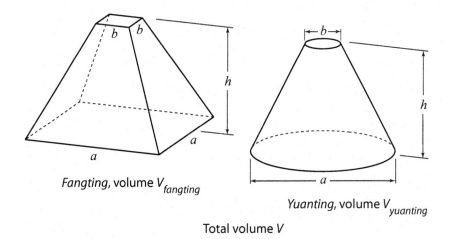

Fangting, volume $V_{fangting}$

Yuanting, volume $V_{yuanting}$

Total volume V

Figure 13. Square and circular storage pits, Problems 13 and 14.

cone). The calculation begins with the calculation of the volume of the *fangting*:

$$V_{fangting} \approx \frac{42V}{75} = 7{,}203 \ chi^3 \qquad (2.19)$$

which is correct with $\pi \approx {}^{22}/_7$. (A derivation of this calculation is given later in the text.) Thereafter the calculation is the same as that of the first part of Problem 7. An intermediate quantity, Area for the Corner *Yangma*, here denoted K_1, is calculated:

$$K_1 = \frac{(a-b)^2}{3} = 147 \ chi^2$$

(using $a-b = [a-h] + [h-b]$) and the equation to be solved is

$$b^3 + \left[(a-b)+(h-b)\right]b^2$$
$$+ \left[(a-b)(h-b)+K_1\right]b = V_{fangting} - K_1\,(h-b)$$
$$b^3 + 35b^2 + 441b = 5{,}145 \ chi^3$$

This has one real root, $b = 7 \ chi$, and the remaining dimensions are

$$h = b + (h–b) = 21 \ chi$$
$$a = h + (a–h) = 28 \ chi$$

The only difference from the first part of Problem 7 is that the relations given there are $a–b$ and $h–b$, while here $a–h$ and $h–b$ are given. The reader is expected to be aware that $a–b = (a–h) + (h–b)$.

◆ *2.1.3.3. Derivation of the volume of the square storage pit*

Following the main text is a comment in smaller characters (Section 3.13.4) which gives a derivation of Equation (2.19). The comment starts with the standard formulas for the volumes of a *fangting* and a *yuanting*, as they are known for example from the *Jiuzhang suanshu* (see Box 4, pp. 38–39), but introducing a named intermediate quantity, the Space,[1] which we denote K_2.

> In the case of a *fangting*, [the sides of] the upper and lower squares [b, a] are multiplied together; further, each is multiplied

---

1 The word *xu* 虛 is used four times in the *Jigu suanjing* to name intermediate quantities (Sections 3.4.3, 3.13.4, 3.14.4, and 3.17.1 below). For convenience we translate it 'Space', which is one of its many possible meanings.

by itself; [the three results] are added and multiplied by the height [h] to make [K_2 =] the Space. Dividing by 3 makes [$V_{fangting}$ =] the volume.

In the case of a *yuanting*, [in the same way] the upper and lower diameters [b, a] are multiplied together; each is multiplied by itself; [the three results] are added and multiplied by the height [h] to make [K_2 =] the Space. Further multiplying by 11 and dividing by 42 makes [$V_{yuanting}$ =] the volume of the *yuanting*.

These calculations are

$$K_2 = \left(ab + a^2 + b^2 \right) h \tag{2.20}$$

$$V_{fangting} = \frac{K_2}{3} \tag{2.21}$$

$$V_{yuanting} = \frac{\pi}{4} \frac{K_2}{3} \approx \frac{11 K_2}{42} \tag{2.22}$$

The task is now to calculate the Space, K_2. The total volume of the two storage pits is

$$V = V_{yuanting} + V_{fangting} = \frac{11 K_2}{42} + \frac{K_2}{3}$$

The comment continues,

In the present case the volumes of a *fang*[*ting*] and a *yuan*[*ting*] are added together. Therefore, when [the total volume, V] is in return multiplied by 42, the result is 11 of the *yuan*[*ting*] Space and 14 of the *fang*[*ting*] Space, and when the whole is divided by 25, one Space volume [K_2] is obtained. Further dividing by 3 gives the volume of the *fangting*.

These calculations are

$$42V \approx 11 K_2 + 14 K_2$$

$$K_2 = \frac{3 \times 4V}{4 + \pi} \approx \frac{42V}{25}$$

$$V_{fangting} = \frac{K_2}{3}$$

And this is in effect a statement of (2.19). The comment then directs the reader to the first part of Problem 7 (Section 2.1.1.1 above):

Therefore the method is sought from the *fangting*. The method is used by which, from the difference between [the sides of] the upper and lower squares [*a–b*], the height difference [*h–b*], and the volume [*V*], [the sides of] the upper and lower squares [*a, b*] and the height [*h*] are calculated. This is why there is a division by the product of 3 and 25.

◆ *2.1.3.4. Problem 14: Six square*
and four circular storage pits

Problem 14 is exactly parallel to Problem 13, even with the same answers. The only difference is that here there are six *fangting* and four *yuanting* instead of one of each. In this case the calculation of the Space, corresponding to (2.19), is

$$V = 26{,}342.4 \; hu \times 2.5 \; chi^2/hu = 65{,}856 \; chi^3$$

$$V_{fangting} = \frac{42V}{384} = 7{,}203 \; chi^3 \tag{2.23}$$

and the rest of the calculation is exactly the same as in Section 2.1.3.2 above.

A comment in smaller characters (Section 3.14.4) follows on from the comment translated immediately above:

Here the multiplication by 42 [gives] 4 times 11 times the Space [K_2] for the circular [pit] and 6 times 14 times the Space [K_2] for the square [pit]; dividing by the sum, [4×11 + 6×14 =] 128, makes one Space volume.

As before, dividing this result by 3 makes the volume of [one] *fangting*. So this is calculated using the method of calculating [the dimensions of] a *fangting* from differences. This is why there is a division by the product of 3 and 128.

$$42V \approx 4 \times 11 K_2 + 6 \times 14 K_2 = 128 K_2$$

$$K_2 = \frac{3V}{4 \times \dfrac{\pi}{4} + 6} \approx \frac{42V}{128} = 21{,}609 \; chi^3$$

2.2. Basic methods

With Problems 7, 9, 13, and 14 we have now introduced a few of Wang Xiaotong's methods.

2.2.1 Similar triangles

An example of Wang Xiaotong's use of similar-triangle considerations can be seen in Problem 7 (Section 2.1.1.2 above, Equation (2.7)). Figure 14 shows a section through DC in Figure 8 (p. 35), perpendicular to $EFGH$. The right triangles CNP and CRS are similar; therefore

$$\frac{c-b}{a-b} = \frac{g}{h}$$

and

$$c = \frac{g(a-b)}{h} + b$$

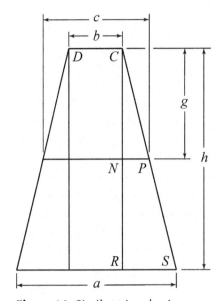

Figure 14. Similar triangles in Problem 7.

Wang Xiaotong uses this kind of consideration often, always in relation to the legs of right triangles, but he never makes it explicit, merely stating the calculation that results.

2.2.2. Volume dissections

The problems discussed in Section 2.1 above, nos. 7, 9, 13, and 14, are examples of a type of problem which occurs several times in the *Jigu suanjing*: The volume of a geometric object is given, together with two or more differences between dimensions of the object, and the dimensions are required. These problems are usually solved by dissecting the object into parts whose volumes are sums of products of known quantities and powers of the dimension which is chosen to be the unknown in a cubic equation.

The simplest example of this method is the first part of Problem 7, discussed in Section 2.1.1.1 above. The dissection is shown in Figure 9 (p. 37), and the dimension chosen as the unknown is the side of the upper square, *b*. The volumes of the parts are given in equations (2.2)–(2.4) (p. 36).

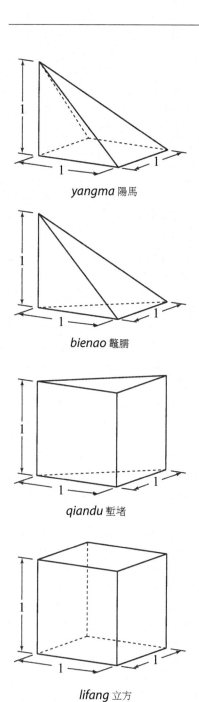

yangma 陽馬

bienao 鱉臑

qiandu 塹堵

lifang 立方

Figure 15. Standard blocks used by Liu Hui in describing volume dissections.

HOW WERE THE DISSECTIONS COMMUNICATED? There is no indication that Wang Xiaotong's book contained diagrams of any sort, nor are diagrams of three-dimensional geometric situations known in other early Chinese mathematical works. If we make the reasonable assumption that the book was intended as an adjunct to a teacher's instruction, then an interesting question is how a teacher communicated these often rather complex geometric constructions to a student. He may perhaps have possessed scale models of each problem, but study of an earlier text suggests a simpler and more flexible means.

The commentary by Liu Hui 劉徽 (3rd century CE) on the *Jiu-zhang suanshu*, chapter 5, describes a number of volume dissections using a set of standard blocks (*qi* 棋, 'chessmen') of four kinds, *yangma*, *bienao*, *qiandu*, and *lifang*, each with dimensions 1 × 1 × 1 *chi* (Figure 15).[1] He describes their placement to form a structure which is equivalent to the geometric structure in question except in its dimensions.

If Wang Xiaotong, or a teacher using his book, possessed a set of such blocks, he could easily have demonstrated the dissection shown in Figure 9, for example, by building a structure which is equivalent in all but its dimensions. It would only be necessary for the student to abstract from the difference in dimensions between the actual geometric situation and the constructed model.

THE REASONING BEHIND THE CALCULATIONS (2.2)–(2.4) is not given in the book, but one can imagine, for example, a teacher expressing (2.2) as follows:

> The total volume of the four corner *yangma* is the product of the Area for the Corner *Yangma* and the height. But the height is the sum of the side of the upper square and the difference between the height and the side of the upper square. Therefore the volume is calculated by: multiplying the Area for the Corner *Yangma* by the difference between the height and the side of the upper square; further multiplying the Area for the Corner *Yangma* by the side of the upper square; and adding the two results together.

(All this verbiage would have been much more concise in the spoken language of its time.) This reasoning shows which calculated quantities belong where in the cubic equation represented on the counting

---

1 Wagner 1979.

board. The only algebraic principle required in the above passage amounts to, given quantities *s*, *t*, and *x*,

$$st = s(t - x) + sx \qquad (2.24)$$

In other parts of the *Jigu suanjing*, to be discussed further below, somewhat more complex algebraic principles appear to be required.

2.2.3. Reasoning about calculations

In this book we have most often avoided the term 'algebra', for there appears to be no consensus among historians of mathematics as to how this word should be applied historically. We occasionally find it convenient to refer to 'algebraic principles', for example (2.24) above, but in general we speak of 'reasoning about calculations'. Some readers will see this as rhetorical algebra, while others will point out that it does not involve manipulation of abstract symbols, and therefore, in their judgement, is not algebra. We shall return to this question in Section 2.10 below.

We have seen a simple example of reasoning about calculations in Section 2.1.1.1 above, where the principle stated in Equation (2.24) is applied in order to express a product as the sum of two products.

A better example is in Problem 7, concerning a square granary (Section 2.1.1.3 above). In the course of deriving the dimensions of the remaining grain the text states the formula for the volume of a *fangting*, then, 'to return to the starting point', manipulates the verbal formula much as we would manipulate an equation.

A more sophisticated piece of reasoning is seen in Problems 13 and 14, discussed in Section 2.1.3 above. Here the total volume of a number of *fangting* and *yuanting* is given, and a necessary first step is to find the volume of one of the *fangting*. The formulas for the volumes of *fangting* and *yuanting* are stated; these are the same as those given in the *Jiuzhang suanshu*, but with the introduction of a name, here translated 'Space', for an intermediate result which is the same in both formulas (Equation (2.20) above). Then the volume of a *fangting* is $^1/_3$ times the Space and the volume of a *yuanting* is $^\pi/_{12} \approx {}^{11}/_{42}$ times the Space ((2.21) and (2.22)). Combining these results with the numbers of *fangting* and *yuanting* in the problem statements gives the result that the total volume would be calculated by multiplying the Space by a known quantity. Since in this case the total volume is known, this calculation can be reversed to give the value of the Space, and dividing this by 3 gives the volume of one of the *fangting*. The method of Problem 7 (Section 2.1.1 above) can then be used to find its dimensions.

Wang Xiaotong often gives names to intermediate quantities in order to simplify the statement of a calculation, but here, in the explanation of an algorithm, he gives a name to an *abstract* quantity, one which, at this point in the text, is unknown. This would seem to represent a new level of abstraction in Chinese mathematical texts.[1]

THE NAMING OF AN UNKNOWN quantity occurs also in a comment in the second part of Problem 7 (Sections 2.1.1 above and 3.7 below). In this way an intermediate result, equivalent to (2.9) (p. 40), is proved. After this comes another form of reasoning about calculations, using area dissection. A dissection, reconstructed in Figure 10 (p. 42), is used to prove two more intermediate results, equivalent to (2.16) and (2.17). Combining these three results proves the calculation of the main text, Equation (2.6) (p. 40).

---

1 In this connection note fn. 1 in Section 2.5.1.2, p. 72 below.

2.3. Problems 10–12: Parallelepipeds and cylinders

2.3.1. Problem 10: A square granary and a circular storage pit

In Problem 10 (Section 3.10 below) the total volume is given for two objects, together with two relations between their dimensions, and the dimensions are required. The objects are the square parallelepiped and cylinder shown in Figure 16; the given quantities are

$$V = 57{,}801.825 \ chi^3$$

$$d\text{–}s = 0.9 \ chi$$

$$s\text{–}h = 29.8 \ chi$$

and the approximation $\pi \approx {}^{22}/_7$ is to be used. The calculation produces the cubic equation,

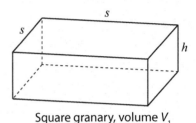

Square granary, volume V_1

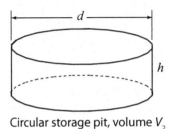

Circular storage pit, volume V_2

Total volume $V = V_1 + V_2$

Figure 16. Square granary and circular storage pit, Problem 10.

$$h^3 + \left[2(d-s)\frac{11}{25} + 2(s-h) \right] h^2$$
$$+ \left\{ \frac{11}{25} \left[2(s-h) + (d-s) \right](d-s) + (s-h)^2 \right\} h = \frac{14}{25} V \qquad (2.25)$$

$$h^3 + 60.392 \ h^2 + 911.998 \ h = 32{,}369.022 \ chi^3$$

This has one real root, $h = 15.5 \ chi$, and the other dimensions are

$$s = (s\text{–}h) + h = 45.3 \ chi$$

$$d = (d\text{–}s) + s = 46.2 \ chi$$

A comment in smaller characters (Section 3.10.4) gives an alternative calculation of the linear coefficient:

$$fangfa = \frac{14(s-h)^2 + 11(d-h)^2}{25} = 911.998 \ chi^2 \qquad (2.26)$$

in which it is necessary to use the identity $d\text{–}h = (d\text{–}s) + (s\text{–}h)$. The comment names this result 'Area for the Corner' (or 'Corners'), which suggests that it may have been derived from the linear coefficient in (2.25) using an area dissection. This dissection might have been as on the right in Figure 17, with 'corner' referring to the squares *MNOP*

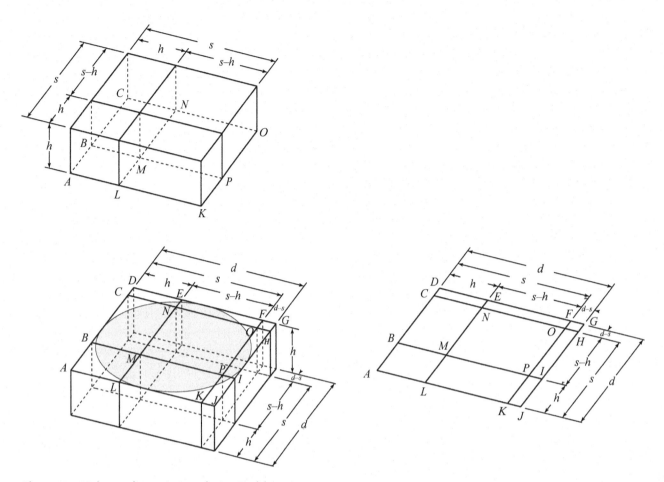

Figure 17. Volume dissection to derive Problem 10.

and *MEGI*, whose areas are $(s-h)^2$ and $(d-h)^2$. On the other hand, we shall note below a simpler alternative equation to (2.25), in which the linear coefficient is equal to (2.26).

BUT THE COMMENT THEN GOES on (Section 3.10.4.2) to give a derivation of the original cubic equation, (2.25), rather than the alternative. The proof uses a volume dissection, which we suggest may be reconstructed as in Figure 17. As we have seen before (Section 2.2.2 above), the dissection divides a known volume into parts whose volumes are products of known quantities and powers of the quantity chosen as the unknown for the cubic equation. The upper object is the square granary, and the lower object is a parallelepiped in which the cylindrical storage pit is inscribed. In the figure the plane *ADGJ* is copied on the right in order to ease reference.

The comment is not as clear and detailed as those discussed in Sections 2.1.1.3 and 2.1.3.3 above, but the basic points are clear enough. The volumes of those parts of the lower object in Figure 17 which have no counterparts in the upper object enter the total volume V multiplied by $\pi/4 \approx {}^{11}/_{14}$. This gives

$$\frac{11}{14}\left\{2(d-s)h^2 + \left[2(s-h)+(d-s)\right](d-s)h\right\} \qquad (2.27)$$

The volumes of those parts which occur in both objects are multiplied by $\pi/4 + 1 \approx {}^{25}/_{14}$, giving

$$\frac{25}{14}\left[h^3 + 2(s-h)h^2 + (s-h)^2 h\right] \qquad (2.28)$$

Multiplying the volume by ${}^{14}/_{25}$ and combining terms gives

$$\begin{aligned}\frac{14V}{25} = h^3 &+ \left[\frac{22}{25}(d-s)+2(s-h)\right]h^2 \\ &+ \left\{\frac{11}{25}\left[2(s-h)+(d-s)\right](d-s)+(s-h)^2\right\}h\end{aligned} \qquad (2.29)$$

which is (2.25).

THE MAIN TEXT THEN GOES on to state the formulas for the volumes of the two objects, given their dimensions:

$$V_1 = s^2 h$$

$$V_2 = \frac{11}{14}d^2 h \approx \frac{\pi}{4}d^2 h$$

These simple formulas had been known since ancient times, and it is not clear why they are repeated here.

♦ *2.3.1.1. Note on a simpler calculation*

We note in passing that a simpler equation, equivalent to (2.25), can be derived by substituting $(d-h) - (s-h)$ for $d-s$ and using (2.26) for the linear coefficient:

$$\begin{aligned}h^3 &+ 2\left[\frac{11}{25}(d-h)+\frac{14}{25}(s-h)\right]h^2 \\ &+ \left[\frac{14}{25}(s-h)^2 + \frac{11}{25}(d-h)^2\right]h = \frac{14}{25}V\end{aligned}$$

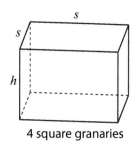

4 square granaries

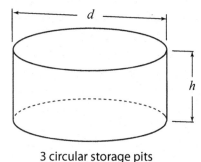

3 circular storage pits

Total volume *V*

Figure 18. Square granaries and circular storage pits, Problem 11.

This can be derived using a dissection which is a slight variation on Figure 17.

2.3.2. Problem 11: Four square granaries and three circular storage pits

Problem 11 is very similar to Problem 10. The total volume is given for *four* square granaries and *three* circular storage pits. See Figure 18. The given quantities are

$$d-s = 10 \ chi$$

$$s-h = 5 \ chi$$

$$V = 40,872 \ chi^3$$

and the value $\pi \approx {}^{22}/_7$ is to be used.

The calculation produces the cubic equation

$$h^3 + \left[2(d-s)\frac{33}{89} + 2(s-h) \right]h^2$$
$$+ \left[\frac{33}{89}\left[2(s-h)+(d-s) \right](d-s)+(s-h)^2 \right]h = \frac{14V}{89} \qquad (2.30)$$

$$h^3 + 17\,{}^{37}/_{89}\ h^2 + 99\,{}^{14}/_{89}\ h = 6,429\,{}^{27}/_{89}\ chi^3 \qquad (2.31)$$

This has one real root, $h = 13 \ chi$, and the other dimensions are

$$s = (s-h) + h = 18 \ chi$$

$$d = (d-s) + s = 28 \ chi$$

A comment in smaller characters refers to the comment in the previous problem (Section 2.3.1 above) and states,

> In this case there are four square granaries, so the 14 is multiplied by 4, and three circular storage pits, so the 11 is multiplied by 3; adding these gives 89, to be divided.

This amounts to saying that the volumes of parts which occur only in the lower object in Figure 17 are to be multiplied by $3\pi/4 \approx {}^{33}/_{14}$, while the volumes of parts which occur in both of the objects are to be multiplied by $3\pi/4 + 4 \approx {}^{89}/_{14}$. Then the counterparts to (2.27) and (2.28) are

$$\frac{33}{14}\left\{ 2(d-s)h^2 + \left[2(s-h)+(d-s) \right](d-s)h \right\} \qquad (2.32)$$

$$\frac{89}{14}\left[h^3+2(s-h)h^2+(s-h)^2h\right] \qquad (2.33)$$

Multiplying by $^{14}/_{89}$ and combining terms gives (2.30).

2.3.3. Problem 12: Another square granary and circular storage pit

Problem 12 is again very like Problem 10, but uses a slightly more complicated derivation of the method. See Figure 19. The total volume of the two objects is given, together with two relations between dimensions,

$$V = 7,680 \ chi^3$$

$$s-g = 2 \ chi$$

$$h-s = 3 \ chi$$

and the value $\pi \approx {}^{22}/_7$ is to be used. The calculation leads to the cubic equation,

$$g^3+\left\{\frac{14\left[(s-g)+(h-s)\right]}{25}+2(s-g)\right\}g^2$$

$$+\left\{\frac{14\times2\left[(s-g)+(h-s)\right](s-g)}{25}+(s-g)^2\right\}g \qquad (2.34)$$

$$=\frac{14V}{25}-\frac{14(s-g)^2\left[(s-g)+(h-s)\right]}{25}$$

$$g^3 + 6.8\ g^2 + 15.2\ g = 4,289.6\ chi^3 \qquad (2.35)$$

This has one real root, $g = 14 \ chi$, and the other dimensions are

$$s = d = g + (s-g) = 16 \ chi$$

$$h = g + (s-g) + (h-s) = 19 \ chi$$

A COMMENT IN SMALLER CHARACTERS (Section 3.12.4) refers to Problem 10 and to a volume dissection which we reconstruct in Figure 20. The object marked I is the square granary and the one marked II is a parallelepiped in which the cylindrical pit is inscribed.

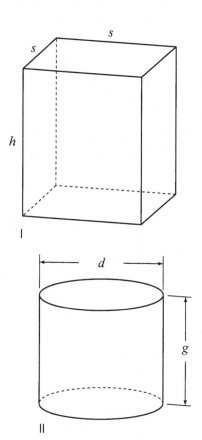

Figure 19. Square granary and circular storage pit, Problem 12.

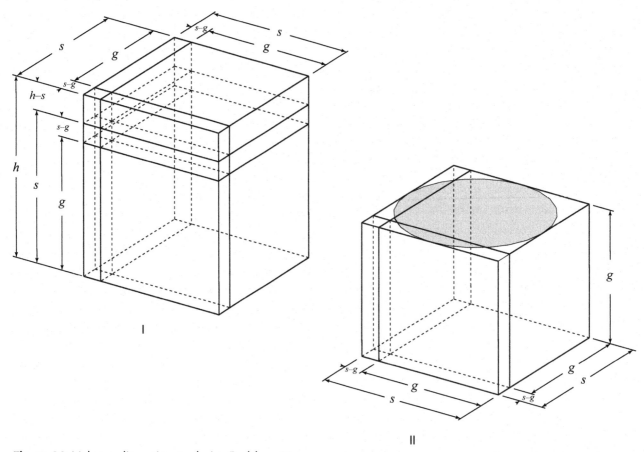

Figure 20. Volume dissection to derive Problem 12.

The volumes of the parts of the dissection which occur in both objects enter into the total volume multiplied by $\pi/4 + 1 \approx {}^{25}/_{14}$:

$$\frac{25}{14}g^3 + 2 \times \frac{25}{14}(s-g)g^2 + \frac{25}{14}(s-g)^2 g$$

and the volumes of the parts which occur only in object I enter unchanged in the total volume:

$$\big[(s-g)+(h-s)\big]g^2 + 2\big[(s-g)+(h-s)\big](s-g)g$$
$$+(s-g)^2\big[(s-g)+(h-s)\big]$$

Adding, multiplying by $^{14}/_{25}$, and combining terms gives

$$\frac{14V}{25} = g^3 + \left\{\frac{14\big[(s-g)+(h-s)\big]}{25} + 2(s-g)\right\}g^2$$

$$+ \left\{\frac{14\times 2\big[(s-g)+(h-s)\big](s-g)}{25} + (s-g)^2\right\}g$$

$$+ \frac{14(s-g)^2\big[(s-g)+(h-s)\big]}{25}$$

which is equivalent to (2.34).

◆ 2.3.3.1. Note on a simpler calculation

We note in passing that a somewhat simpler equation than (2.34) can be formed by substituting h–g for $(s$–$g) + (h$–$s)$:

$$g^3 + \left\{\frac{14(h-g)}{25} + 2(s-g)\right\}g^2$$

$$+ \left\{\frac{14\times 2(h-g)(s-g)}{25} + (s-g)^2\right\}g$$

$$= \frac{14V}{25} - \frac{14(s-g)^2(h-g)}{25}$$

This equation could be derived by a simpler dissection, with Figure 21 substituted for the left side of Figure 20.

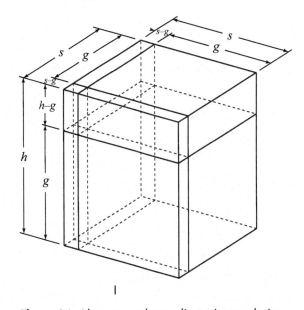

Figure 21. Alternate volume dissection to derive Problem 12. Compare Figure 20, left side.

2.4. Problem 8: A wedge

In this discussion Problem 8 (Section 3.8 below) is the first which can be said to reflect a practical situation. An earthwork is to be constructed of tamped earth by corvée labourers from two counties; the volumes of earth to be moved and tamped by each group are given, and the dimensions of the parts constructed by each are required. The official in charge could presumably then order each of the two counties in turn to build to the calculated dimensions. There is good historical evidence that this kind of problem of apportioning labour was commonly encountered by Chinese officials involved in public works (see Section 1.2 above), though in real life the calculations would no doubt have been more complicated than this, with many more variables to be taken into account as well as non-quantitative considerations.

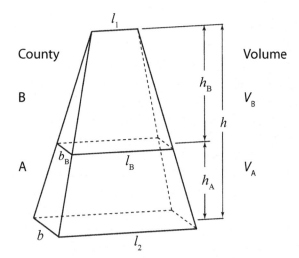

Figure 22. *Chumeng*, right rectangular wedge, Problem 8.

2.4.1. The problem

The dimensions of a *chumeng* 芻甍 (a right rectangular wedge) are given, as shown in Figure 22:

$$l_1 = 30 \ chi$$
$$l_2 = 90 \ chi$$
$$b = 60 \ chi$$
$$h = 120 \ chi$$

This object is divided into two parts by a plane parallel to the base, and the volume of the upper part is

$$V_B = 243 \text{ persons} \times 36 \ chi^3/\text{person·day}$$
$$\times 8 \text{ days} = 69{,}984 \ chi^3$$

The height and dimensions of each part (h_A, h_B, b_B, l_B) are to be calculated.

2.4.2. The calculation

The equation whose root is found is

$$h_B{}^3 + \frac{3K_1}{2} h_B{}^2 = \frac{6V_B h^2}{2(l_2 - l_1)b} \tag{2.36}$$

where K_1, called 'Height for the Upper Length', is

$$K_1 = \frac{hl_1}{l_2 - l_1} = 60 \; chi \tag{2.37}$$

Thus

$$h_B{}^3 + 90h_B{}^2 = 839{,}808 \; chi^3$$

This has one real root, $h_B = 72 \; chi$. The other height is

$$h_A = h{-}h_B = 48 \; chi$$

and the other dimensions are found by considering similar triangles:

$$b_B = \frac{h_B}{h} b = 36 \; chi \tag{2.38}$$

$$l_B = \frac{h_B}{h}(l_2 - l_1) + l_1 = 66 \; chi \tag{2.39}$$

2.4.3. Derivation of the calculation

A comment in smaller characters (Section 3.8.4) gives a derivation of (2.36). It first states the formula for the volume of a *chumeng*, which had been known since the *Jiuzhang suanshu* (see Box 4, p. 38):

> This 'B volume' [V_B] was originally [obtained by] doubling the lower length [l_B], adding the upper length [l_1], multiplying by the lower breadth [b_B] and the height [h_B], and dividing by 6; this makes the volume of a [*chu*]*meng*.

This amounts to

$$V_B = \frac{(2l_B + l_1)b_B h_B}{6} \tag{2.40}$$

The comment continues,

> Here, returning to the origin, this is multiplied by 6 and by the area of [the square on] the height [h^2] . . . and divided by the difference in lengths [$l_2{-}l_1$] multiplied by the breadth [b].

This calculation is,

$$\frac{6V_B h^2}{(l_2 - l_1)b}$$

which is twice the constant term in (2.36). Then,

> The result is the assigned height multiplied by itself $[h_B^2]$ multiplied by 3 times the Height for the Upper Length $[K_1]$.

In modern terminology,

$$3K_1 h_B^2$$

Further,

> Therefore three of the smaller height $[K_1]$ is the *lianfa* [the quadratic term of the equation].

This, $3K_1$, is twice the quadratic term of (2.36). Finally,

> There remain two cubes made of the assigned height $[h_B]$.

The result obtained in this comment is thus

$$\frac{6V_B h^2}{(l_2 - l_1)b} = 2h_B^3 + 3K_1 h_B^2 \tag{2.41}$$

which is twice (2.36). The comment then states that the coefficients are to be halved. The comment ends here; it gives no hint of the reasoning behind (2.41).

THE MENTION OF 'CUBES' ABOVE suggests that a derivation by volume dissection might be intended, but we have been unable to discover one. A modern algebraic derivation of (2.41) is as follows. It will be seen that this would be difficult to express without the use of modern symbolism.

From (2.40),

$$\frac{6V_B h^2}{b(l_2 - l_1)} = \frac{(2l_B + l_1)b_B h_B h^2}{b(l_2 - l_1)} \tag{2.42}$$

Considering similar triangles, as in (2.38) and (2.39),

$$\frac{h}{b} = \frac{h_B}{b_B} \tag{2.43}$$

$$\frac{l_B}{l_1} = \frac{h_B}{h}\frac{l_2 - l_1}{l_1} + 1 = \frac{h_B}{K_1} + 1 \tag{2.44}$$

Substituting first (2.43), then (2.37), and then (2.44) into (2.42),

$$\frac{6V_B h^2}{b(l_2 - l_1)} = \frac{(2l_B + l_1)h_B^2 h}{(l_2 - l_1)}$$

$$= \frac{(2l_B + l_1)h_B^2 K_1}{l_1} = \left(\frac{2l_B}{l_1} + 1\right)h_B^2 K_1$$

$$= \left(\frac{2h_B}{K_1} + 2\right)h_B^2 K_1 + h_B^2 K_1$$

$$= 2h_B^3 + 3K_1 h_B^2$$

which is (2.41).

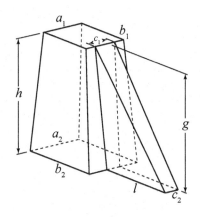

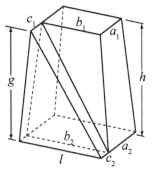

Figure 23. Grand astrologer's platform and ramp, Problem 2: two possible interpretations.

2.5. Problems 2 and 6: The Grand Astrologer's platform and ramp, and a related problem

2.5.1. Problem 2: The platform and ramp[1]

In Problem 2 (Section 3.2 below), a tamped-earth platform and associated ramp are described, to be built by workers from two counties, A and B. Two possible interpretations of the description are shown in Figure 23. Other interpretations are possible, but the mathematics is the same whichever interpretation is followed.

The basic geometric situation is shown in Figure 24. The following quantities are given:

$$a_2 - a_1 = 20 \ chi$$
$$b_2 - b_1 = 40 \ chi$$
$$b_1 - a_1 = 30 \ chi$$
$$h - a_1 = 110 \ chi \quad\quad (2.45)$$
$$c_1 - c_2 = 12 \ chi$$
$$l - c_1 = 104 \ chi$$
$$g - l = 40 \ chi$$

Labour data is given which allows the straightforward calculation of the volumes of the two structures, V and W, and the volumes of the contributions of the two counties, V_A, V_B, W_A, and W_B:

$$V = 1{,}740{,}000 \ chi^3$$
$$V_A = 531{,}750 \ chi^3$$
$$V_B = 1{,}208{,}250 \ chi^3$$
$$W = 352{,}800 \ chi^3$$
$$W_A = 81{,}900 \ chi^3$$
$$W_B = 270{,}900 \ chi^3$$

---

1 On these problems, see also Li Yan 1963d: 78–80; 1998: 128–138; Qian Baocong 1966a; Bréard 1999: 95–99, 353–356. Andrea Bréard appears to have been the first to note that the names given by Wang Xiaotong to intermediate quantities give clues to how the calculations were derived.

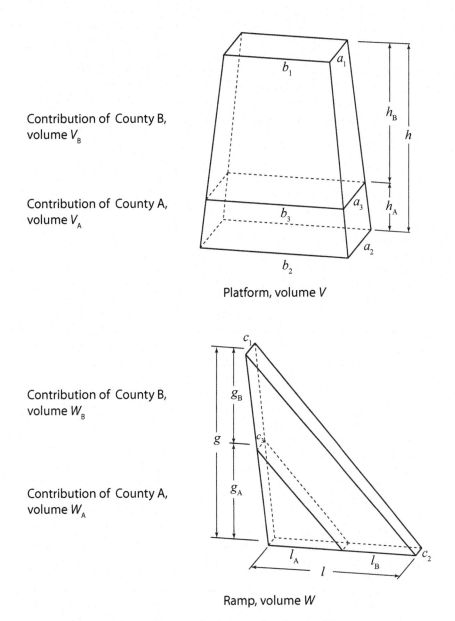

Contribution of County B, volume V_B

Contribution of County A, volume V_A

Platform, volume V

Contribution of County B, volume W_B

Contribution of County A, volume W_A

Ramp, volume W

Figure 24. Grand Astrologer's platform and ramp, Problem 2: contributions of the counties.

The dimensions of the two structures and of the two counties' contributions are required.

It will be seen that the first two parts of Problem 2 resemble the two parts of Problem 7 (Section 2.1.1 above). The difference is that the granary of Problem 7 is square, while the platform of Problem 2 is rectangular (Figures 8 and 24, pp. 35 and 65). Therefore Problem 2 has five equations in five unknowns, while Problem 7 has only three equations in three unknowns.

◆ *2.5.1.1. The dimensions of the platform*

The first task is to calculate a_1 in Figure 24.

Wang Xiaotong's calculation proceeds as follows (Section 3.2.3.1 below). He names five intermediate quantities:

Area for the Corner *Yangma*,

$$K_1 = \frac{(a_2 - a_1)(b_2 - b_1)}{3} = 266^2/_3 \, chi^2$$

Truncated Volume for the Corner *Yangma*,

$$K_2 = K_1(h - a_1) = 29{,}333^1/_3$$

Area for the Corner Heads,

$$K_3 = \frac{(a_2 - a_1)(b_1 - a_1)}{2} = 300 \, chi^2$$

Truncated Volume for the Corner Heads,

$$K_4 = K_3(h - a_1) = 33{,}000 \, chi^3$$

Determined Number,

$$K_5 = \frac{(a_2 - a_1) + (b_2 - b_1)}{2} = 30 \, chi$$

The 'areas' and 'volumes' here do not correspond to any area or volume in the geometric situation: these are simply arbitrary names for intermediate quantities.

The cubic equation whose root is to be extracted is then

$$a_1^3 + \left[(h - a_1) + (b_1 - a_1) + K_5 \right] a_1^2$$

$$+ \left\{ \left[K_5 + (b_1 - a_1) \right](h - a_1) + K_1 + K_3 \right\} a_1 \qquad (2.46)$$

$$= V - (K_2 + K_4)$$

$$a_1^3 + 170 a_1^2 + 7{,}166^2/_3 \, a_1 = 1{,}677{,}666^2/_3 \, chi^3$$

This equation has one real root, $a_1 = 70 \, chi$. The other dimensions are then

$$a_2 = (a_2 - a_1) + a_1 = 90 \ chi$$
$$b_1 = (b_1 - a_1) + a_1 = 100 \ chi$$
$$b_2 = (b_2 - b_1) + b_1 = 140 \ chi$$
$$h = (h - a_1) + a_1 = 180 \ chi$$

In Figures 23–24 we have shown the platform as a frustum of a right wedge, but the calculation is correct for a frustum of any rectangular wedge, and the names given to intermediate quantities in the calculation suggest that Wang Xiaotong may have derived (2.46) using a volume dissection similar to that shown in Figure 25, in which one edge of the frustum is perpendicular to the base.

The platform is divided into parts whose volumes are sums of products of known quantities and powers of the unknown, a_1. The 'corner heads' are nos. I and II, with volumes,

$$V_\mathrm{I} = \frac{(a_2 - a_1)b_1 h}{2}$$
$$= \frac{1}{2}\big[(a_2 - a_1)(b_1 - a_1)(h - a_1) + (a_2 - a_1)(b_1 - a_1)a_1 \quad (2.47)$$
$$+ (a_2 - a_1)(h - a_1)a_1 + (a_2 - a_1)a_1^2\big]$$

$$V_\mathrm{II} = \frac{a_1(b_2 - b_1)h}{2}$$
$$= \frac{1}{2}\big[a_1(b_2 - b_1)(h - a_1) + a_1^2(b_2 - b_1)\big] \quad (2.48)$$

$$V_\mathrm{I} + V_\mathrm{II} = K_5 a_1^2 + K_5(h - a_1)a_1 + K_3 a_1 + K_4$$

The 'corner *yangma*' is no. III, with volume,

$$V_\mathrm{III} = \frac{(a_2 - a_1)(b_2 - b_1)h}{3}$$
$$= \frac{1}{3}\big[(a_2 - a_1)(b_2 - b_1)(h - a_1) + (a_2 - a_1)(b_2 - b_1)a_1\big] \quad (2.49)$$

$$= K_1 a_1 + K_2$$

And no. IV has volume,

$$V_\mathrm{IV} = a_1 b_1 h$$
$$= a_1(b_1 - a_1)(h - a_1) + a_1^2(b_1 - a_1) \quad (2.50)$$
$$+ a_1^2(h - a_1) + a_1^3$$

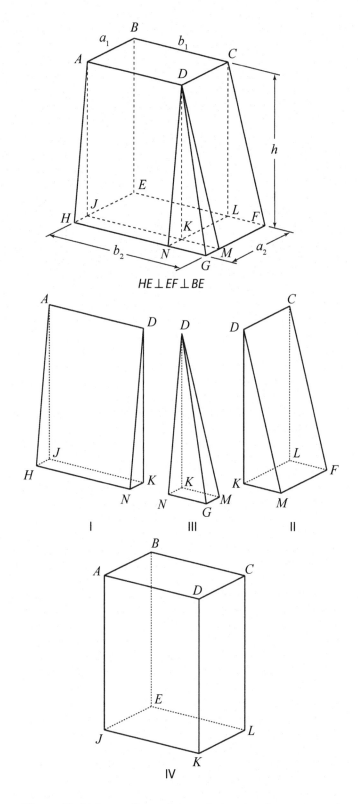

$HE \perp EF \perp BE$

Figure 25. Volume dissection to derive the dimensions of the platform.

Summing,

$$V = V_{\mathrm{I}} + V_{\mathrm{II}} + V_{\mathrm{III}} + V_{\mathrm{IV}}$$

$$= a_1^3 + \left[(h - a_1) + (b_1 - a_1) + K_5 \right] a_1^2$$

$$+ \left\{ \left[K_5 + (b_1 - a_1) \right] (h - a_1) + K_1 + K_3 \right\} a_1 \; + \; K_2 \; + \; K_4$$

which is equivalent to (2.46).

In Section 2.2.2 above, Equation (2.24) (p. 51), a simple algebraic principle was stated which can be used to derive (2.48) and (2.49). Derivation of (2.47) and (2.50) requires in each case either application of (2.24) three times, or a more complicated algebraic principle: given quantities s, t, and x,

$$st = (s{-}x)(t{-}x) + (t{-}x)x + (s{-}x)x + x^2$$

which can be demonstrated by dissection of a rectangle with dimensions s and t.

◆ *2.5.1.2. The contributions of the*
two counties to the platform

The next task (Section 3.2.3.2) is to calculate h_{B} in Figure 24. Here again Wang Xiaotong gives names to quantities that will be used more than once:

Height for the Upper Width,

$$K_6 = \frac{ha_1}{a_2 - a_1} = 630 \; chi \tag{2.51}$$

Height for the Upper Length,

$$K_7 = \frac{hb_1}{b_2 - b_1} = 450 \; chi \tag{2.52}$$

These 'heights' are not equal to any dimensions in the geometric situation; they are simply intermediate results in the calculation.

The height h_{B} of county B's contribution is then a root of the equation

$$h_{\mathrm{B}}^3 + \tfrac{3}{2}(K_6 + K_7)h_{\mathrm{B}}^2 + 3K_6 K_7 h_{\mathrm{B}} = \frac{3h^2 V_{\mathrm{B}}}{(a_2 - a_1)(b_2 - b_1)} \tag{2.53}$$

$$h_B{}^3 + 1{,}620 h_B{}^2 + 850{,}500 h_B = 146{,}802{,}375 \; chi^3$$

This equation has one real root, $h_B = 135 \; chi$. The remaining dimensions are then calculated by considering similar triangles:

$$a_3 = \frac{(a_2 - a_1) h_B}{h} + a_1 = 85 \; chi$$

$$b_3 = \frac{(b_2 - b_1) h_B}{h} + b_1 = 130 \; chi$$

A comment in smaller characters (Section 3.2.3.3) gives a remarkable explanation of (2.53). The text of the comment is corrupt as it has come down to us. Luo Tengfeng (1770–1841)[1] has proposed an interpretation and a large number of emendations to the text, and Qian Baocong follows most of these in his edition. We have no doubt that his analysis of the underlying mathematics is correct, even if his precise emendations may be less certainly correct. In the translation we follow all of his emendations.

The comment uses some geometric designations for intermediate quantities, but the reasoning concerns calculations rather than geometry, and thus might be considered almost purely algebraic.

The platform and the contributions of the two counties (see Figure 26) are what is known in classical Chinese mathematics as *chutong* 芻童, frusta of a rectangular wedge. The calculation of the volume in the *Jiuzhang suanshu* (see Box 4, p. 38) is

$$V_B = \frac{h_B \left[(2a_1 + a_3) b_1 + (2a_3 + a_1) b_3 \right]}{6}$$

$$= \frac{h_B (2a_1 b_1 + a_3 b_1 + 2a_3 b_3 + a_1 b_3)}{6}$$

The formula is not stated in the comment but clearly is used. Substituting this expression for V_B into (2.53) gives

$$h_B{}^3 + \frac{3(K_6 + K_7)}{2} h_B{}^2 + 3 K_6 K_7$$

$$= \frac{h^2}{2(a_2 - a_1)(b_2 - b_1)} h_B (2a_1 b_1 + a_3 b_1 + 2a_3 b_3 + a_1 b_3) \tag{2.54}$$

Doubling this gives

1 Luo Tengfeng 1993, 1: 16a–18b.

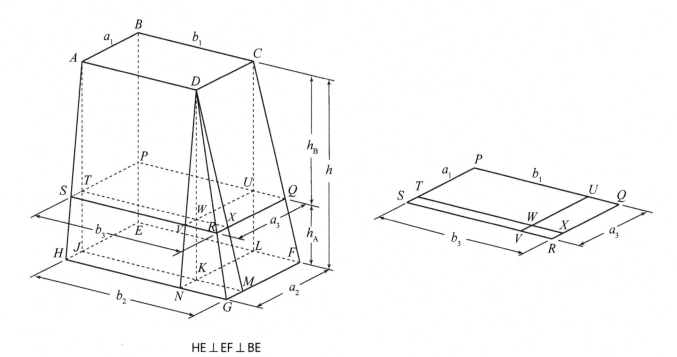

HE ⊥ EF ⊥ BE

Figure 26. Dissection of the platform to derive the contributions of the counties.

$$2h_{\mathrm{B}}^{3} + 3\left(K_{6} + K_{7}\right)h_{\mathrm{B}}^{2} + 6K_{6}K_{7}h_{\mathrm{B}}$$
$$= \frac{h^{2}}{\left(a_{2} - a_{1}\right)\left(b_{2} - b_{1}\right)} h_{\mathrm{B}}\left(2a_{1}b_{1} + a_{3}b_{1} + 2a_{3}b_{3} + a_{1}b_{3}\right) \quad (2.55)$$

The four summed terms in parentheses on the right side of (2.55) are the areas of the rectangles $PUWT$, $PUVS$, $PQRS$, and $PQXT$ respectively in Figure 26. The comment relates the left side of (2.55) to the four areas.

The comment first states that multiplying the Height for the Upper Width by the Height for the Upper Length (K_{6} and K_{7}, see (2.51) and (2.52)) 'gives two of the small area'. (In fact, twice this product gives this result.) The 'small area' is $PUWT$ in Figure 26, with area $a_{1}b_{1}$:

$$2K_{6}K_{7} = 2\frac{h^{2}}{\left(a_{2} - a_{1}\right)\left(b_{2} - b_{1}\right)}a_{1}b_{1} \quad (2.56)$$

The linear term of (2.55) is $6K_{6}K_{7}h_{\mathrm{B}}$. Equation (2.56) indicates that two of these six $K_{6}K_{7}$'s correspond to the term $2a_{1}b_{1}$ in (2.55); four remain to be accounted for.

The comment introduces two more named quantities,

Height for the Lower Length,

$$K_8 = \frac{hb_3}{b_2 - b_1} \qquad (2.57)$$

Height for the Lower Width,

$$K_9 = \frac{ha_3}{a_2 - a_1} \qquad (2.58)$$

These quantities correspond to no dimensions in the geometric situation, and at this point their numeric values are unknown; they are abstract quantities which figure in the algebraic reasoning which follows. They are not actually defined in the commentary; presumably the reader is expected to deduce their definitions by analogy with the Heights for the Upper Length and Width (see equations (2.51) and (2.52) above).[1]

The comment states, 'multiplying the Height for the Lower Length [K_8] by the Height for the Upper Width [K_6] gives one of the middle areas'. The 'middle areas' are $PQXT$ (area a_1b_3) and $PUVS$ (area a_3b_1). In this case,

$$K_6K_8 = \frac{h^2}{(a_2 - a_1)(b_2 - b_1)} a_1 b_3$$

This is followed by a statement that

$$K_8 = K_7 + h_B$$

$$K_9 = K_6 + h_B$$

which is true because, considering similar triangles,

$$\frac{h_B}{h} = \frac{a_3 - a_1}{a_2 - a_1} = \frac{b_3 - b_1}{b_2 - b_1}$$

so that

$$K_7 + h_B = h\frac{b_1}{b_2 - b_1} + h\frac{b_3 - b_1}{b_2 - b_1} = \frac{hb_3}{b_2 - b_1} = K_8$$

---

1   In an earlier publication (Lim & Wagner 2013a: 21) concerning this problem we stated that Wang Xiaotong was unable or unwilling to name unknown quantities. Study of Problem 7 (Section 2.1.1 above) indicates that this is not entirely correct.

$$K_6 + h_{\mathrm{B}} = h\frac{a_1}{a_2 - a_1} + h\frac{a_3 - a_1}{a_2 - a_1} = \frac{ha_3}{a_2 - a_1} = K_9$$

Then 'the middle area [*KLPQ*] is composed of one small area [*PUWT*] and also an area obtained by multiplying the Height for the Upper Width [K_6] by the truncated height [h_{B}]':

$$\frac{h^2}{(a_2 - a_1)(b_2 - b_1)}a_1 b_3 = K_6 K_7 + K_6 h_{\mathrm{B}}$$

This has accounted for one more of the $K_6 K_7$'s and one of the K_6's on the left side of (2.55).

Further, 'multiplying [. . .] by the Height for the Upper Length [K_7] gives one of the middle areas'. The other multiplicand is missing in the text, but the context indicates that it can only be the Height for the Lower Width (K_9), so that the 'middle area' is *PUVS*:

$$K_7 K_9 = \frac{h^2}{(a_2 - a_1)(b_2 - b_1)}a_3 b_1$$

and 'within the middle area there is one small area [*PUWT*, see (2.56)] together with the area obtained by multiplying the Height for the Upper Length [K_7] by the truncated height [h_{B}]':

$$\frac{h^2}{(a_2 - a_1)(b_2 - b_1)}a_3 b_1 = K_7 K_9 = K_6 K_7 + K_7 h_{\mathrm{B}}$$

which accounts for two more of the $K_6 K_7$'s and one of the K_7's on the left side of (2.55).

Finally, 'multiplying the Height for the Lower Width [K_9] by the Height for the Lower Length [K_8] gives two of the large area'. The 'large area' is *PQRS*, with area $a_3 b_3$:

$$K_8 K_9 = \frac{h^2}{(a_2 - a_1)(b_2 - b_1)}a_3 b_3$$

and 'within the large area [*PQRS*] there is one small area [*PUWT*], and in addition there are the Heights for the Upper Width and the Upper Length [K_6, K_7], each multiplied by the truncated height [h_{B}] to make one middle area, and the truncated height [h_{B}] multiplied by itself to make one area':

$$\begin{aligned} K_8 K_9 &= (K_6 + h_{\mathrm{B}})(K_7 + h_{\mathrm{B}}) \\ &= K_6 K_7 + K_6 h_{\mathrm{B}} + K_7 h_{\mathrm{B}} + h_{\mathrm{B}}^2 \end{aligned}$$

Or,

$$\frac{2h^2}{(a_2 - a_1)(b_2 - b_1)} a_3 b_3$$

$$= 2K_6 K_7 + 2K_6 h_B + 2K_7 h_B + 2h_B^2$$

which accounts for the rest of the terms on the left side of (2.55).

In conclusion, 'Thus there are two of the area formed by multiplying the truncated height by itself [h_B^2] and six of the small area [$K_6 K_7$]. There are also three each of the Heights for the Upper Width and the Upper Length [K_6, K_7].' The quantities mentioned here are

$$2h_B^2$$
$$6K_6 K_7$$
$$3K_6 h_B$$
$$3K_7 h_B$$

The text continues,

> These, multiplied by the truncated height [h_B], give the six are-as.[1] When all are halved, the small area [the term $K_6 K_7$] is multiplied by three. Further, there are three each of the Heights for the Upper Width and the Upper Length [K_6, K_7]. When [the terms $3K_6 h_B$ and $3K_7 h_B$ in (2.59)] are halved once, in each case one and one half is obtained; this is why [these terms] are multiplied by three and divided by two. All of the areas are multiplied by the truncated height [h_B] to make volumes in [cubic] *chi*.
>
> $$2h_B^2 + 6K_6 K_7 + 3K_6 h_B + 3K_7 h_B$$
> $$= \frac{h^2}{(a_2 - a_1)(b_2 - b_1)} (2a_1 b_1 + a3b_1 + 2a_3 b_3 + a_1 b_3) \qquad (2.59)$$

Halving Equation (2.59) and multiplying by h_B gives (2.54), which is equivalent to (2.53).

◆ *2.5.1.3. The dimensions of the ramp*

The next task (Section 3.2.3.4) is to calculate c_2 in Figures 23 and 24 (pp. 64 and 65). Names given for quantities in the calculation are:

---

1 Note fn. 1 in Section 3.2.3.3, p. 136.

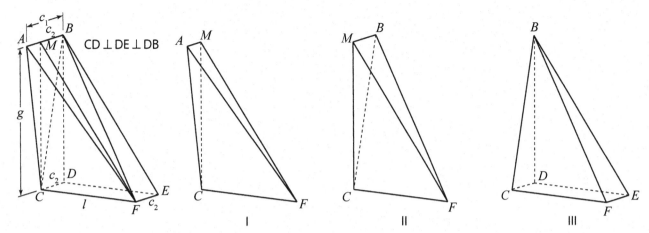

Figure 27. Dissection to derive the dimensions of the ramp.

Area for the Corner *Yangma*,

$$K_{11} = (g - c_2)(l - c_2) = 18,096 \ chi^2$$

Volume for the *Bienao* Corner,

$$K_{12} = K_{11}(c_1 - c_2) = 217,152 \ chi^3$$

Area for the Vertical and Horizontal Edge *Bienao*,

$$K_{13} = ([l - c_2] + [g - c_2])(c_1 - c_2) = 3,264 \ chi^2$$

The cubic equation whose root is to be extracted is then

$$c_2^3 + \left[\frac{c_1 - c_2}{3} + (l - c_2) + g - c_2\right]c_2^2 + \left(\frac{K_{13}}{3} + K_{11}\right)c_2 \qquad (2.60)$$
$$= \frac{6W - K_{12}}{3}$$

$$c_2{}^3 + 276 \ c_2{}^2 + 19,184 \ c_2 = 633,216 \ chi^3$$

This equation has one real root, $c_2 = 24 \ chi$. The remaining dimensions of the ramp are then

$$c_1 = c_2 + (c_1 - c_2) = 36 \ chi$$
$$l = c_1 + (l - c_1) = 140 \ chi$$
$$g = l + (g - l) = 180 \ chi$$

The names given to intermediate quantities suggest that (2.60) was derived using a dissection similar to that shown in Figure 27. The volume is divided into parts whose volumes are sums of products of known quantities and powers of c_2. The 'vertical and horizontal edge *bienao*' is no. I, with volume

$$W_{\mathrm{I}} = \frac{gl(c_1 - c_2)}{6}$$

$$= \frac{1}{6}\left[(g-c_2)(l-c_2)(c_1-c_2) + (g-c_2)(c_1-c_2)c_2\right.$$

$$\left. + (l-c_2)(c_1-c_2)c_2 + (c_1-c_2)c_2^2\right]$$

$$= \frac{(c_1-c_2)c_2^2 + K_{13}c_2 + K_{12}}{6}$$

The '*bienao* corner' is no. II, with volume

$$W_{\mathrm{II}} = \frac{glc_2}{6}$$

$$= \frac{1}{6}\left[(g-c_2)(l-c_2)c_2 + (g-c_2)c_2^2 + (l-c_2)c_2^2 + c_2^3\right]$$

$$= \frac{c_2^3 + (l-c_2)c_2^2 + (g-c_2)c_2^2 + K_{11}c_2}{6}$$

The 'corner *yangma*' is no. III, with volume

$$W_{\mathrm{III}} = \frac{glc_2}{3} = 2W_{\mathrm{II}}$$

Combining gives the volume of the ramp:

$$W = W_{\mathrm{I}} + W_{\mathrm{II}} + W_{\mathrm{III}} = W_{\mathrm{I}} + 3W_{\mathrm{II}}$$

$$6W = K_{12} + K_{13}c_2 + (c_1-c_2)c_2^2 + 3K_{11}c_2$$

$$+ 3(g-c_2)c_2^2 + 3c_2^2 + 3c_2^3$$

$$= 3c_2^3 + \left[3(g-c_2) + 3(l-c_2) + (c_1-c_2)\right]c_2^2$$

$$+ (3K_{11} + K_{13})c_2 + K_{12}$$

And from this (2.60) can easily be derived.

◆ *2.5.1.4. The contributions of the two counties to the ramp*

The final task (Section 3.2.3.5) is to calculate l_A in Figure 24. The cubic equation whose root is to be extracted is

$$l_A^3 + \frac{3c_2 l}{c_1 - c_2} l_A^2 = \frac{6 W_A l^2}{(c_1 - c_2) g} \qquad (2.61)$$

$$l_A^3 + 840\, l_A^2 = 4{,}459{,}000 \ chi^3$$

This equation has one positive root, $l_A = 70 \ chi$. The remaining dimensions are then

$$c_3 = c_2 + \frac{c_1 - c_2}{l} l_A = 30 \ chi$$

$$g_A = \frac{g}{l} l_A = 90 \ chi$$

In this case Wang Xiaotong gives us no clue as to how he derived his calculation, but the dissection shown in Figure 28 will serve. A necessary implicit assumption is that *NPQR* is parallel to *ABEF*. (Figure 28 is drawn with $ABDC \perp CDEF \perp BDE$, but this is not a necessary condition.) The contribution of county A is *NPQRCD*. Dissecting this into the *qiandu SPQRCD* and the *bienao NSRC* gives for its volume

$$W_A = \frac{g_A l_A c_2}{2} + \frac{g_A l_A (c_3 - c_2)}{6}$$

Considering similar triangles,

$$\frac{l_A}{l} = \frac{g_A}{g} = \frac{c_3 - c_2}{c_1 - c_2}$$

whence,

$$W_A = \frac{g c_2}{2l} l_A^2 + \frac{g(c_1 - c_2)}{6 l^2} l_A^3$$

and multiplying by $6 l^2 / g(c_1 - c_2)$ gives (2.61).

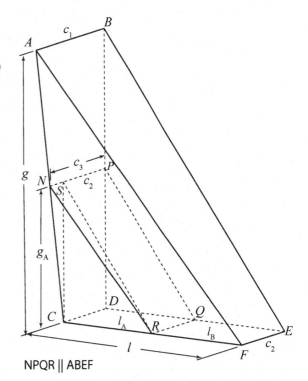

NPQR ∥ ABEF

Figure 28. Dissection of the ramp to derive the contributions of the counties.

2.5.2. Problem 6: A rectangular grain cellar

Problem 6 (Section 3.6 below) is equivalent to the first half of Problem 2 (Sections 2.5.1.1–2.5.1.3 above), with an interesting difference which will be noted further below.

The volume of a cellar with the same form as the astrologer's platform (a *chumeng*, see Box 4, p. 38) is given, together with four relations between its dimensions, and the dimensions are required. Then the volume of one county's contribution to the digging of the cellar is given, and the dimensions of this part are required.

◆ *2.5.2.1. The dimensions of the cellar*

See Figure 29. The given quantities are

$$V = 285,000 \; chi^3$$
$$b_1 - a_1 = 10 \; chi$$
$$b_2 - b_1 = 30 \; chi$$
$$b_1 - h = 60 \; chi \qquad (2.62)$$
$$a_2 - b_1 = 10 \; chi$$

The equation whose root is to be extracted is

$$h^3 + \left[\frac{(a_2 - a_1) + (b_2 - b_1)}{2} + K_3 \right] h^2$$
$$+ \left[\tfrac{1}{2}(b_2 - b_1)(a_1 - h) + K_1 + K_2 \right] h = V \qquad (2.63)$$

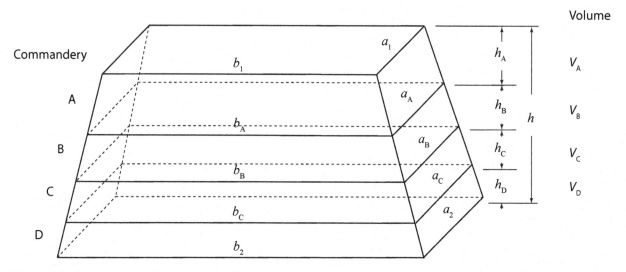

Figure 29. Rectangular grain cellar, Problem 6.

where

$$K_1 = \frac{(a_2 - a_1)(b_2 - b_1)}{3} = 200 \ chi^2$$

$$K_2 = \left[(a_1 - h) + \tfrac{1}{2}(a_2 - a_1) \right] (b_1 - h) = 3600 \ chi^2$$

$$K_3 = (b_1 - h) + (a_1 - h) = 110 \ chi$$

Numerically,

$$h^3 + 135 \ h^2 + 4550 \ h = 285,000 \ chi^3 \tag{2.64}$$

This has one real root, $h = 30 \ chi$, and the other dimensions are

$$b_1 = (b_1 - h) + h = 90 \ chi$$
$$a_1 = (a_1 - h) + h = (b_1 - h) - (b_1 - a_1) + h = 80 \ chi$$
$$b_2 = (b_2 - h) + h = (b_1 - h) + (b_2 - b_1) + h = 120 \ chi$$
$$a_2 = (a_2 - h) + h = (a_2 - b_1) + (b_1 - h) + h = 100 \ chi$$

Equation (2.63) can be derived by the same volume dissection as was shown for (2.46) in Problem 2, Figure 25. The intermediate calculations which correspond to (2.47)–(2.50) are

$$V_I = \frac{b_1(a_2 - a_1)h}{2} = \frac{(a_2 - a_1)(b_1 - h)}{2}h + \frac{(a_2 - a_1)}{2}h^2$$

$$V_{II} = \frac{a_1(b_2 - b_1)h}{2} = \frac{(b_2 - b_1)(a_1 - h)}{2}h + \frac{(b_2 - b_1)}{2}h^2$$

$$V_{III} = \frac{(a_2 - a_1)(b_2 - b_1)}{3}h$$

$$V_{IV} = a_1 b_1 h = \left[a_1(b_1 - h) + a_1 h \right] h$$

$$= (a_1 - h)(b_1 - h)h + \left[(b_1 - h) + (a_1 - h) \right] h^2 + h^3$$

- *2.5.2.1.1. Avoiding negative coefficients*

The important difference between this problem and the first part of Problem 2 is that a direct application of the method of Problem 2 to

Problem 6 would necessitate working with negative numbers. Using (2.46) from Section 2.5.1.1 above (p. 66) would give

$$a_1{}^3 - 15\,a_1{}^2 - 1450\,a_1 = 300,000 \; chi^3$$

This has one real root, $a_1 = 80$ *chi*, which is correct. Negative numbers were used in China as early as the *Jiuzhang suanshu*, and in times later than Wang Xiaotong polynomials occur with negative coefficients, but negative numbers occur nowhere in Wang Xiaotong's book. He always chooses the dimension to be solved for in such a way that negative coefficients are avoided.

We do not know whether calculators in Wang Xiaotong's time were able to find roots of polynomials with negative coefficients, but even if they could it would still have been wise to avoid them, for they open the possibility of multiple positive roots. For example, if in Problem 6 we imagine a very different shape for the grain cellar of Figure 29, substituting the quantity

$$a_1 - b_1 = 326$$

for $b_1 - a_1$ in (2.62), and we use the method of Problem 2, then (2.46) is, numerically,

$$a_1{}^3 - 855a_1{}^2 + 229,382a_1 = 18,947,328 \; chi^3$$

This has three positive real roots, $a_1 \approx 161.1$, 294.0, and 399.8 *chi*. Only this last would be a solution to the problem; the others imply that some dimensions are negative. On the other hand, if (2.63) is used, then the equation to be solved would be

$$h^3 + 303h^2 + 16,310h = 285,000$$

This has only one positive real root, $h \approx 13.8$ *chi*, and the calculator need not worry about false roots. But the two methods give the same result.

◆ 2.5.2.2. The contribution of County A

As in the second part of Problem 2, the volume of the upper part of the object (A in Figure 29) is given, and its dimensions are required. The calculation is exactly the same as in Problem 2, resulting in the numerical equation

$$h_A{}^3 + 315\, h_A{}^2 + 32{,}400\, h_A = 435{,}888 \; chi^3$$

This has one real root, $h_A = 12 \; chi$, and the other dimensions are

$$b_A = \frac{h_A\left(b_2 - b_1\right)}{h} + b_1 \;\; = 102 \; chi$$

$$a_A = \frac{h_A\left(a_2 - a_1\right)}{h} + a_1 \;\; = 88 \; chi$$

2.6. Problems 3 and 5: Construction of a dyke, and a related problem

2.6.1. Problem 3: the dyke

Problem 3 (Section 3.3 below) concerns the object shown in Figure 30, a tamped-earth dyke. Its volume is given, together with five relations between dimensions, and the dimensions are required. The dyke is constructed by workers from four counties. The volume of each county's contribution is given, and the dimensions of each contribution are required. (On this problem note also Qian Baocong 1966a.)

It can be seen in Figure 30 that there are seven unknown dimensions to be calculated in the first part of the problem; but only six equations are given. However, an implicit assumption becomes apparent in the calculation, namely $a_1 = a_2$. In particular, there are two points in the text (pp. 144 and 145) in which the 'upper width' is mentioned without an indication of whether the upper width of the eastern or western end of the dyke is meant, i.e., whether a_1 or a_2 is meant. We denote this quantity a, and so there are only six unknowns, a, b_1, b_2, h_1, h_2, and l (s is easily calculated from these).

The given quantities are

$$V = 275,924,800 \ cun^3$$
$$b_2 - a_2 = 682 \ cun$$
$$b_1 - a_1 = 62 \ cun$$

ABCD ⊥ CDEF ⊥ EFGH

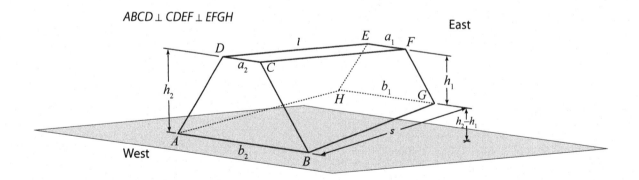

Figure 30. Dyke, Problem 3.

$$h_2 - h_1 = 310 \; cun \qquad\qquad (2.65)$$

$$a - h_1 = 49 \; cun$$

$$l - h_1 = 4769 \; cun$$

◆ *2.6.1.1. The dimensions of the dyke*

The calculation gives names to four intermediate results:

Area for the *Bienao*,

$$K_1 = \frac{\left(h_2 - h_1\right)\left(b_2 - b_1\right)}{6}$$

$$= \frac{\left(h_2 - h_1\right)\left[\left(b_2 - a_2\right) - \left(b_1 - a_1\right)\right]}{6} = 32{,}033\tfrac{1}{3} \; cun^2 \quad (2.66)$$

Area for the End of the Large Recumbent *Qiandu*,

$$K_2 = \frac{\left(h_2 - h_1\right)\left(b_1 - a_1\right)}{2} = 9{,}610 \; cun^2$$

Area for the End of the Small Recumbent *Qiandu*,

$$K_3 = \frac{\left(h_2 - h_1\right)\left(a - h_1\right)}{2} = 7{,}595 \; cun^2$$

Proportion for the Large and Small *Qiandu* and *Bienao*,

$$K_4 = K_1 + K_2 + K_3 = 49{,}238\tfrac{1}{3} \; cun^2$$

The cubic equation whose root is to be extracted is then

$$h_1^{\,3} + \left[\left(l - h_1\right) + \left(a - h_1\right) + \frac{h_2 - h_1}{2} + \frac{b_1 - a_1}{2}\right] h_1^{\,2}$$

$$+ \left\{\left[\frac{h_2 - h_1}{2} + \frac{b_1 - a_1}{2} + \left(a - h_1\right)\right]\left(l - h_1\right) + K_4\right\} h_1 \qquad (2.67)$$

$$= V - K_4\left(l - h_1\right)$$

$$h_1^{\,3} + 5{,}004 \; h_1^{\,2} + 1{,}169{,}953\tfrac{1}{3} \; h_1 = 41{,}107{,}188\tfrac{1}{3} \; cun^3$$

This has one positive root, $h_1 = 31 \; cun$, and the other dimensions are

$$a = a_1 = a_2 = h_1 + (a - h_1) = 80 \; cun$$

$$b_1 = a + (b_1 - a_1) = 142 \ cun$$

$$b_2 = a + (b_2 - a_2) = 762 \ cun$$

$$h_2 = h_1 + (h_2 - h_1) = 341 \ cun$$

$$l = h_1 + (l - h_1) = 4,800 \ cun$$

$$s = \sqrt{l^2 + \left(h_2 - h_1\right)^2} = 4,810 \ cun \tag{2.68}$$

(This last calculation indicates that $ABCD \perp CDEF \perp EFGH$.)

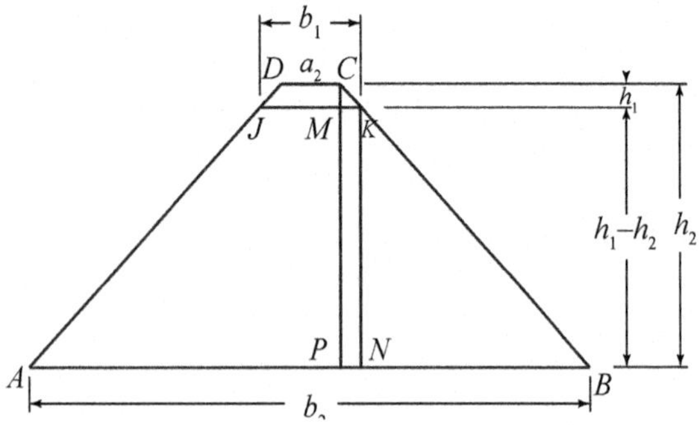

Figure 31. Western wall of the dyke, Problem 3.

• *2.6.1.1.1. Note on a simpler calculation*

We note in passing that this calculation is unnecessarily complicated. Given the implicit assumption that $a_1 = a_2$, considering similar triangles alone,

$$h_1 = \frac{\left(b_1 - a_2\right)\left(h_2 - h_1\right)}{b_2 - b_1}$$

$$= \frac{\left(b_1 - a\right)\left(h_2 - h_1\right)}{\left(b_1 - a\right) - \left(b_2 - a\right)} = 31 \ cun$$

Figure 31 shows the western wall of the dyke, dissected by the lines JK, CP, and KN. The similar triangles are CMK and KNB.

♦ *2.6.1.2. Derivation of the calculation of the dimensions*

The names given to intermediate results indicate that (2.67) was derived using a volume dissection like that shown in Figure 32. The volume of each part can be expressed as a sum of products of known quantities and powers of h_1.

The '*bienao*' are the two parts marked I, each with volume

$$V_I = \frac{K_1 l}{2} = \frac{K_1\left(l - h_1\right)}{2} + \frac{K_1 h_1}{2}$$

The 'large recumbent *qiandu*' are the two parts marked II, each with volume

$$V_{II} = \frac{K_2 l}{2} = \frac{K_2\left(l - h_1\right)}{2} + \frac{K_2 h_1}{2}$$

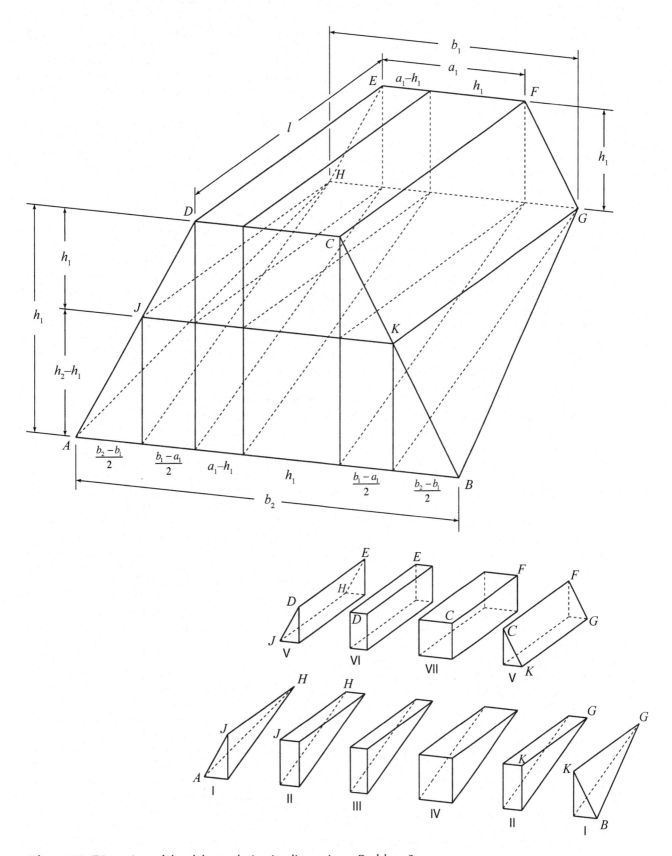

Figure 32. Dissection of the dyke to derive its dimensions, Problem 3.

And the 'small recumbent *qiandu*' is the part marked III, with volume

$$V_{\mathrm{III}} = K_3 l = K_3 \left(l - h_1 \right) + K_3 h_1$$

The other parts have volumes

$$V_{\mathrm{IV}} = \frac{l \left(h_2 - h_1 \right)}{2} h_1 = \frac{\left(l - h_1 \right) \left(h_2 - h_1 \right)}{2} h_1 + \frac{h_2 - h_1}{2} h_1^2$$

$$V_{\mathrm{V}} = \frac{l \left(b_1 - a_1 \right)}{4} h_1 = \frac{\left(l - h_1 \right) \left(b_1 - a_1 \right)}{4} h_1 + \frac{\left(b_1 - a_1 \right)}{4} h_1^2$$

$$V_{\mathrm{VI}} = l \left(a_1 - h_1 \right) h_1 = \left(l - h_1 \right) \left(a_1 - h_1 \right) h_1 + \left(a_1 - h_1 \right) h_1^2$$

$$V_{\mathrm{VII}} = l h_1^2 = \left(l - h_1 \right) h_1^2 + h_1^3$$

Then

$$V = 2 V_{\mathrm{I}} + 2 V_{\mathrm{II}} + V_{\mathrm{III}} + V_{\mathrm{IV}} + 2 V_{\mathrm{V}} + V_{\mathrm{VI}} + V_{\mathrm{VII}}$$

$$= K_1 \left(l - h_1 \right) + K_2 \left(l - h_1 \right) + K_3 \left(l - h_1 \right) + K_4 h_1$$

$$+ \frac{\left(l - h_1 \right) \left(h_2 - h_1 \right)}{2} h_1 + \frac{\left(l - h_1 \right) \left(b_1 - a_1 \right)}{2} h_1$$

$$+ \left(l - h_1 \right) \left(a_1 - h_1 \right) h_1 + \frac{h_2 - h_1}{2} h_1^2 + \frac{b_1 - a_1}{2} h_1^2$$

$$+ \left(a_1 - h_1 \right) h_1^2 + \left(l - h_1 \right) h_1^2 + h_1^3$$

and this is equivalent to (2.67).

◆ *2.6.1.3. Implicit assumptions*

We noted in Section 2.6.1 above that with seven unknowns and six equations, an additional assumption is required, and that there is good reason to assume that the implicit assumption is $a_1 = a_2$. In particular, the difference $b_2 - b_1$ is used once (see Equation (2.66)), and this cannot be calculated from the given quantities (2.65) without an assumption of some relation between a_1 and a_2.

There is, however, another possible interpretation, which we present without making any claim about its relation to Wang Xiaotong's original intentions. Both of the mentions of the 'upper width' without indication of whether a_1 or a_2 is meant occur directly before a mention of the eastern end of the dyke. It might therefore be reasonable to assume that a_1 is meant, so that, in (2.65), we should write $a_1-h_1 = 49$ *cun*. If then the missing assumption is taken to be

$$b_2-b_1 = 620 \; cun$$

and no assumption is made about the relation between a_1 and a_2, then Equation (2.67) gives the correct answers, including $a_1 = a_2 = 80$ *cun*.

Though we find this interpretation interesting, we note that if $a_1 \neq a_2$ a derivation of (2.67) by dissection would be more complicated than the one we have suggested in Figure 32.

◆ 2.6.1.4. The contributions of the counties

The parts of the dyke constructed by the four counties are shown in Figure 33. County A is first, and starts at the eastern end. The text gives the calculation of the dimensions for county A, then states that the others are calculated in the same way.

The volume of County A's contribution is

$$V_A = 4{,}960 \; cun^3 \, / \, man \times 6{,}724 \text{ men} = 33{,}351{,}040 \; cun^3$$

and the dimensions l_A, b_{2A}, h_A, and s_A are required. The text gives names to two intermediate results:

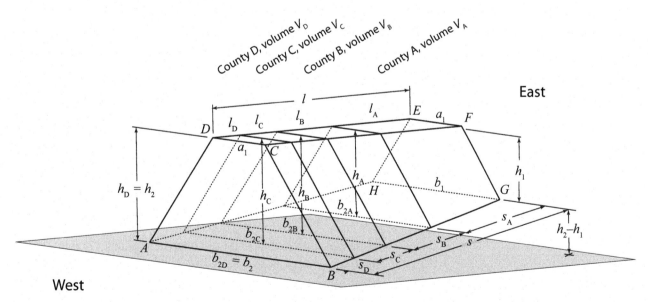

Figure 33. Contributions of the counties, Problem 3.

Divisor,

$$K_5 = (b_2-b_1)(h_2-h_1) = 192,200 \ cun^2$$

Area for the End Wall,

$$K_6 = 3(a_1+b_1)h_1 = 20,646 \ cun^2$$

The 'end wall' is presumably $EFGH$, whose area is $\frac{1}{2}(a_1+b_1)h_1$.
Then the cubic equation whose root is to be extracted is

$$l_A^3 + \frac{3b_1 l}{b_2-b_1}l_A^2 + \frac{K_6 l^2}{K_5}l_A = \frac{6V_A l^2}{K_5} \tag{2.69}$$

$$l_A^3 + 3,298 \, ^2/_{31} \, l_A^2 + 2,474,941 \, ^{29}/_{31} \, l_A$$
$$= 23,987,761,548 \, ^{12}/_{31} \ cun^3$$

This has one real root, $l_A = 1,920 \ cun$, and the other dimensions are,
considering similar triangles,

$$b_{2A} = \frac{l_A(b_2-b_1)}{l} + b_1 = 390 \ cun$$

$$s_A = \sqrt{l_A^2 + (h_A - h_1)^2} = 1924 \ cun \tag{2.70}$$

$$h_A = \frac{(h_2-h_1)l_A}{l} + h_1 = 155 \ cun$$

◆ *2.6.1.5. Derivation of the contribution of County A*

A comment in smaller characters (Section 3.3.3.4) gives a derivation
for (2.69). The text of the comment is clearly corrupt, and commentators have suggested numerous emendations to try to make sense
of it. In the translation we discuss the various emendations, but do
not attempt to reconstruct the original wording. Here we suggest a
plausible reconstruction of the original *intention* of the comment.

The beginning of the comment is unproblematic:

> In this case there is a level dyke at the top and a *yanchu* at
> the bottom. The difference between the two heights is then
> the height of the *[yan]chu*. The *[yan]chu* has on each side a
> *bienao* and in the middle a *qiandu*.

Elsewhere in the book, volume dissections are merely implied by the names given for intermediate results, but this comment is a clear description of the dissection shown in Figure 34. The 'level dyke' is *DCKJHEFG*, the *yanchu* is *ABKJHG*, the two *bienao* are *AJLH* and *BMKG*, and the *qiandu* is *JKMLHG*. The contribution of County A consists of the 'level dyke' *NPRQHEFG*, the *qiandu QRUTHG*, and the two *bienao STQH* and *UVRG*.

We define

$$V_{\text{A level dyke}} = \text{volume of } NPRQHEFG = \tfrac{1}{2}(a_1 + b_1)h_1 l_A = \frac{K_6 l_A}{6}$$

$$V_{\text{A qiandu}} = \text{volume of } QRUTHG = \tfrac{1}{2}b_1(h_A - h_1)l_A$$

$$V_{\text{A bienao}} = \text{sum of the volumes of } STQH \text{ and } UVRG$$

$$= \tfrac{1}{6}(b_{2A} - b_1)(h_A - h_1)l_A$$

Considering similar triangles,

$$b_{2A} - b_1 = (b_2 - b_1)\frac{l_A}{l}$$

$$h_A - h_1 = (h_2 - h_1)\frac{l_A}{l}$$

$$V_{\text{A qiandu}} = \frac{b_1(h_2 - h_1)}{2l}l_A{}^2 = \frac{b_1 K_5}{2l(b_2 - b_1)}l_A{}^2$$

$$V_{\text{A bienao}} = \frac{(b_2 - b_1)(h_2 - h_1)}{6l^2}l_A{}^3 = \frac{K_5}{6l^2}l_A{}^3$$

Then

$$V_A = V_{\text{A bienao}} + V_{\text{A qiandu}} + V_{\text{A level dyke}}$$

$$= \frac{K_5}{6l^2}l_A{}^3 + \frac{b_1 K_5}{2l(b_2 - b_1)}l_A{}^2 + \frac{K_6}{6}l_A \qquad (2.71)$$

which is equivalent to (2.69).

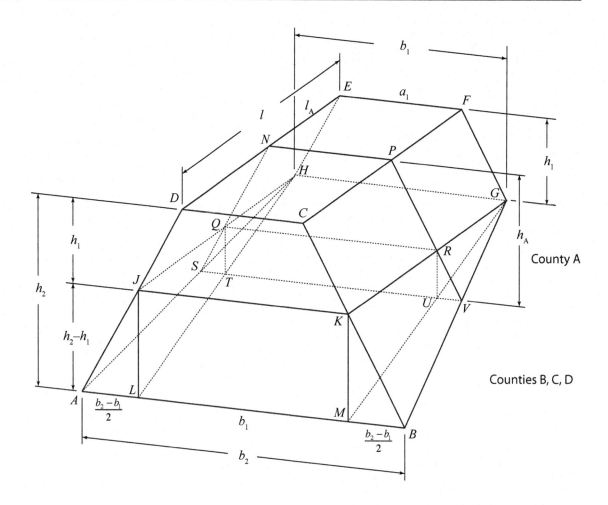

Figure 34. Dissection to derive the contributions of the counties, Problem 3.

2.6.2. Calculation of the volume of a dyke

The main text concludes (Section 3.3.3.5) with a formula for the volume of a geometric object like the dyke in Figure 30, equivalent to

$$V = \frac{1}{6} l \left[\left(2h_2 + h_1\right)\frac{a_2 + b_2}{2} + \left(2h_1 + h_2\right)\frac{a_1 + b_1}{2} \right] \qquad (2.72)$$

While the foregoing text has implicitly assumed that $a_1 = a_2 > h_1$, this equation is valid for arbitrary dimensions of the dyke.

Shen Kangshen (1964) reviews attempts by several pre-modern mathematicians to reconstruct the reasoning behind this formula. The first to give a plausible reconstruction was Jiang Weizhong 蔣維鍾 in 1899: The dyke can be dissected into two *yanchu* in two ways, shown in Figure 35. The volume of a *yanchu* is treated in the *Jiuzhang suanshu*, where a correct formula is given (see Box 4, p. 39). The

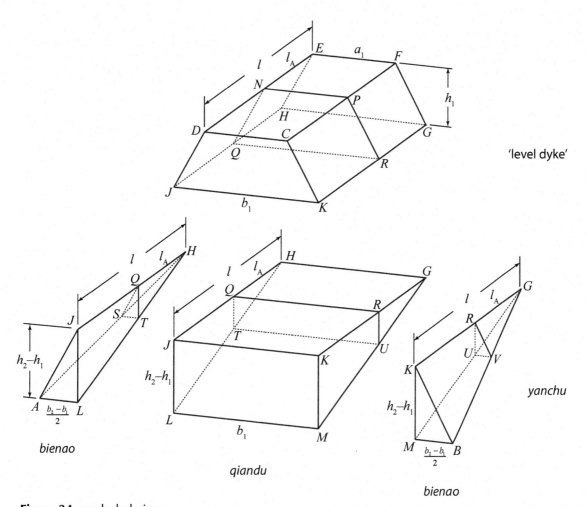

Figure 34, exploded view

two dissections give different results for the volume of the dyke, each of which is correct but neither of which is equivalent to (2.72):

$$V_a = \frac{(a_1 + a_2 + b_2)lh_2}{6} + \frac{(a_1 + b_1 + b_2)lh_1}{6} \qquad (2.73)$$

$$V_b = \frac{(a_2 + b_1 + b_2)lh_2}{6} + \frac{(a_1 + a_2 + b_1)lh_1}{6} \qquad (2.74)$$

These results differ because the method does not take into account the facts that in Figure 30, $BCFG$ and $ADEH$ are planes and $ABCD \parallel EFGH$, so that $\angle EFG = \angle DCB$ and $\angle CDA = \angle FEH$. Given these angle equalities, we have

$$\frac{b_2 - a_2}{h_2} = \frac{b_1 - a_1}{h_1}$$

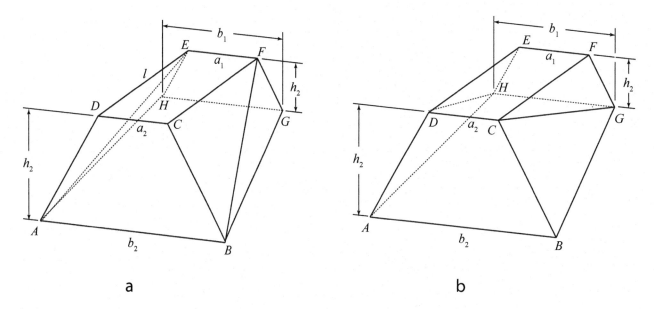

a   b

Figure 35. Two possible dissections to derive the volume of a dyke, Problem 3.

Substituting this equality into either (2.73) or (2.74) gives (2.72).

Instead Jiang Weizhong states incorrectly that the two formulas are approximations, with V_a too large and V_b too small. He therefore takes their average,

$$V = \frac{V_a + V_b}{2}$$

and this is indeed equivalent to Wang Xiaotong's formula, (2.72). As Shen Kangshen writes, Jiang Weizhong's reconstruction is plausible, since Wang Xiaotong constantly uses dissections of solids, and often shows himself capable of the kind of algebraic reasoning required.

Shen Kangshen has, however, a very attractive alternative suggestion. It uses Cavalieri's Theorem, which Chinese historians today call 'Zu Xuan's Axiom'. Zu Xuan 祖暅 (also called Zu Geng or Zu Gengzhi 祖暅之) in the late 5th century CE used this axiom in a derivation of the volume of a sphere (see e.g. Wagner 1978b; Lam Lay Yong 1985), and Wang Xiaotong mentions him and his (now lost) book, *Zhui shu* 綴術, in his preface (Section 3.0.3 below).

See Figure 36. The solid on the right, $A'B'C'D'E'F'G'H'$, is placed so that $A'B'C'D'$ lies in the same plane as $ABCD$. Then any plane parallel to $ABCD$ intersects the two solids in equal areas. By Zu Xuan's Axiom, the volumes of the two solids are equal. The solid on the right, a frustum of a rectangular wedge, is called a *chutong* 芻

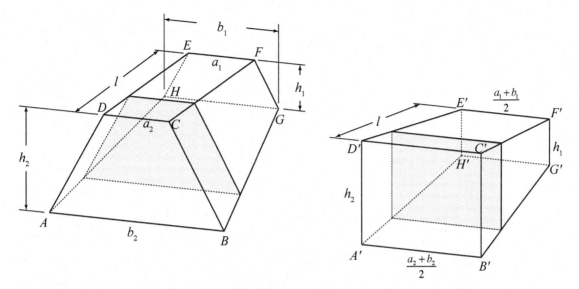

Figure 36. Alternate derivation of the volume of a dyke, Problem 3.

董 in classical Chinese mathematics, and a correct formula for its volume is given in the *Jiuzhang suanshu* (see Box 4, p. 38):

$$V = \frac{1}{6}l\left[(2h_2 + h_1)\frac{a_2 + b_2}{2} + (2h_1 + h_2)\frac{a_1 + b_1}{2}\right]$$

which is identical to (2.72).

2.6.3. Problem 5: A canal and a canal bank

Problem 5 (Section 3.5 below) concerns the objects shown in Figures 37 and 38, a canal and a tamped-earth canal bank made from the earth excavated from the canal. Because of the tamping the volume of the bank is $^3/_4$ of the volume of the canal.

The dimensions of the canal are given, along with the volumes of the contributions of four commanderies in its construction. The dimensions of the contributions are required.

For the bank, the volume and all of the dimensions except v_1 are given, and v_1 is required.

The given quantities are:

$l = 3{,}456$ *chi*

$a = 37.2$ *chi*

$h_1 = 18.6$ *chi*

$$h_2 = 241.8 \; chi$$

$$b_1 = 74.4 \; chi$$

$$b_2 = 520.8 \; chi$$

$$g = 223.2 \; chi$$

$$v_2 = 406.705 \; chi$$

$$V_A = 10,627,891.2 \; chi^3$$

$$V = 67,254,624 \; chi^3$$

The calculations concerning the contributions of the commanderies are equivalent to the calculations in Section 2.6.1.4 above. In Figures 37 and 38 we use the same letter symbols as in Figures 30 and 33 in order to make the equivalence clear. The length of Commandery A's contribution is a root of

$$l_A^3 + \frac{3b_1 l}{b_2 - b_1} l_A^2 + \frac{K_3 l^2}{K_2} l_A = \frac{6 V_A l^2}{K_2}$$

in which

$$K_2 = (b_2 - b_1)(h_2 - h_1) = 99,636.48 \; chi^2$$

$$K_3 = 3(a + b_1)h_1 = 6,227.28 \; chi$$

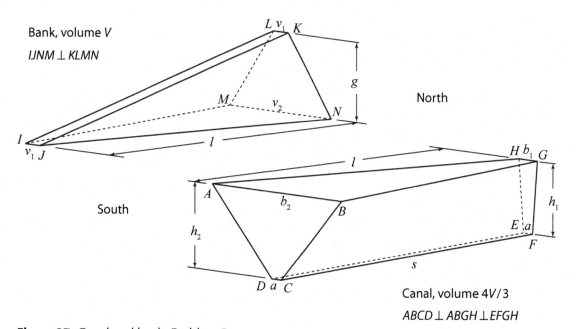

Figure 37. Canal and bank, Problem 5.

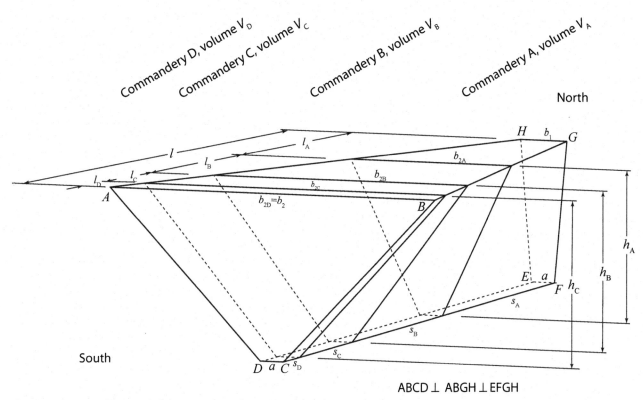

Figure 38. Contributions of the commanderies, Problem 5.

Numerically,

$$l_A^3 + 1728l_A^2 + 746,496l_A = 7,644,119,040 \ chi^3$$

This has one real root, $l_A = 1440 \ chi$, and the other dimensions are

$$b_{2A} = \frac{(b_2 - b_1)l_A}{l} + b_1 = 260.4 \ chi$$

$$h_A = \frac{(h_2 - h_1)l_A}{l} + h_1 = 111.6 \ chi$$

$$s_A = \sqrt{l_A^2 + (h_A - h_1)^2} = 1443 \ chi$$

The text then states that the dimensions of the other commanderies'
contributions are calculated in the same way.

THE CALCULATION GIVEN FOR THE upper width of the bank is

$$v_1 = \frac{1}{2}\left(\frac{6V}{gl} - v_2\right) = 58.21 \ chi \qquad (2.75)$$

The bank is a *yanchu*, whose volume is given in the *Jiuzhang suanshu* (see Box 4, p. 39) as

$$V = \frac{1}{6}(2v_1 + v_2)gl$$

and (2.75) follows directly from this.

2.7. Problem 4: Construction of a 'dragon tail' dyke

2.7.1. The problem

Problem 4 concerns another kind of tamped-earth dyke, the object shown in Figure 39. Its volume is given, together with three relations between dimensions, and the dimensions are required. The dyke is constructed by workers from three counties. The volume of each county's contribution is given, and the dimensions of each contribution are required.

The first two parts of the problem are equivalent to the two parts of the Grand Astrologer's ramp, Sections 2.5.1.3 and 2.5.1.4 above. Figure 39 uses the same letter symbols as Figure 27 (p. 75) in order to make the comparison obvious. The third part of the problem is equivalent to one part of Problem 3, Section 2.6.1.4 above.

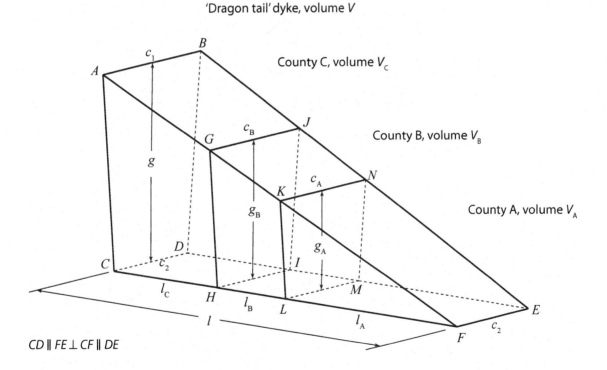

$CD \parallel FE \perp CF \parallel DE$

$ABDC \parallel GHIJ \parallel KLMN \perp CDEF$

Figure 39. 'Dragon tail' dyke, Problem 4.

In Problem 4 the given quantities are

$$V = 19,800 \ chi^3$$
$$c_1 - c_2 = 6 \ chi$$
$$g - c_2 = 12 \ chi$$
$$l - c_2 = 48 \ chi$$

2.7.2. The dimensions of the dyke

In the calculation of the dimensions the following intermediate results are named, using names closely corresponding to those in Problem 2, Section 2.5.1.3 above:

Space Volume,[1]

$$K_1 = 6V = 118,800 \ chi^3$$

Area for the Corner,[2]

$$K_2 = (g-c_2)(l-c_2) = 576 \ chi^2$$

Volume for the Corner *Bienao*,[3]

$$K_3 = K_2(c_1-c_2) = 3,456 \ chi^3$$

Area for the *Bienao* Crossing to the Side,[4]

$$K_4 = \left[(g-c_2) + (l-c_2) \right](c_1 - c_2) = 360 \ chi^2$$

The equation whose root is to be extracted is then

$$c_2{}^3 + \left[\frac{c_1-c_2}{3} + (l-c_2) + (g-c_2) \right]c_2{}^2$$
$$+ \left(\frac{K_4}{3} + K_2 \right)c_2 = \frac{K_1 - K_3}{3} \tag{2.76}$$

which is equivalent to (2.60) in Section 2.5.1.3 above. Numerically,

$$c_2{}^3 + 62c_2{}^3 + 696c_2 = 38,448 \ chi^3$$

---

1 Problem 2: 'Volume'.

2 Problem 2: 'Area for the Corner *Yangma*', K_{11}.

3 Problem 2: same, K_{12}.

4 Problem 2: 'Area for the Vertical and Horizontal Edge *Bienao*', K_{13}.

This has one real root, $c_2 = 18$ *chi*, and the other dimensions are

$$g = (g-c_2)+c_2 = 30 \ chi$$
$$c_1 = (c_1-c_2)+c_2 = 24 \ chi$$
$$l = (l-c_2)+c_2 = 60 \ chi$$

2.7.3. The contribution of County A

The volumes of the contributions of the three counties are

$$V_A = 4{,}702.5 \ chi^3$$
$$V_B = 4{,}708.44 \ chi^3$$
$$V_C = 10{,}389.06 \ chi^3$$

And indeed $V = V_A + V_B + V_C$. The calculation of the dimensions of the contribution of County A is exactly the same as the corresponding calculation concerning the ramp in Problem 2, Section 2.5.1.4 above. The equation whose root is to be extracted is

$$l_A{}^3 + \frac{3c_2 l}{c_1 - c_2} l_A{}^2 = \frac{6V_A l^2}{(c_1 - c_2)g}$$

which is identical to (2.61) in Section 2.5.1.4. Numerically,

$$l_A{}^3 + 594 \ l_A{}^2 = 682{,}803 \ chi^3$$

This has one positive root, $l_A = 33$ *chi*. The other dimensions are, considering similar triangles,

$$g_A = \frac{l_A g}{l} = 15 \ chi$$

$$c_A = \frac{l_A(c_1 - c_2)}{l} + c_2 = 21 \ chi$$

2.7.4. The contribution of County B

The calculation of the dimensions of County B's contribution is equivalent to the calculation in Problem 3, Section 2.6.1.4 above, of the first contribution to the building of a dyke.

The equation whose root is to be extracted is

$$l_{\mathrm{B}}^{3} + \frac{3c_{\mathrm{A}}(l-l_{\mathrm{A}})}{c_1 - c_{\mathrm{A}}}l_{\mathrm{B}}^{2} + \frac{3g_{\mathrm{A}}(c_{\mathrm{A}}+c_2)(l-l_{\mathrm{A}})^2}{(c_1-c_{\mathrm{A}})(g-g_{\mathrm{A}})}l_{\mathrm{B}}$$

$$= \frac{6V_{\mathrm{B}}(l-l_{\mathrm{A}})^2}{(c_1-c_{\mathrm{A}})(g-g_{\mathrm{A}})}$$

(2.77)

which is equivalent to (2.69) in Section 2.6.1.4. Numerically,

$$l_{\mathrm{B}}^{3} + 693l_{\mathrm{B}}^{2} + 42{,}471l_{\mathrm{B}} = 683{,}665.488 \ chi^3$$

This has one positive root, $l_{\mathrm{B}} = 13.2 \ chi$. The other dimensions (not explicitly calculated in the text) are

$$g_{\mathrm{B}} = \frac{l_{\mathrm{B}}g}{l} + g_{\mathrm{A}} = 21 \ chi$$

$$c_{\mathrm{A}} = \frac{g_{\mathrm{B}}(c_1-c_2)}{g} + c_2 = 22.2 \ chi$$

2.7.5. An obscure comment

A comment in smaller characters is very difficult to understand. In our translation (Section 3.4.4.4 below) we have very tentatively

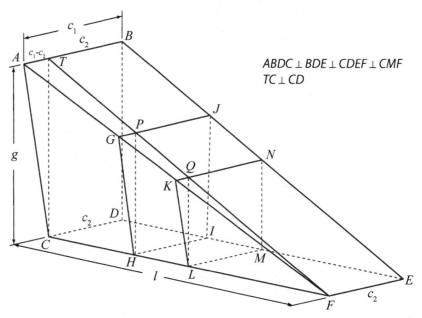

ABDC ⊥ BDE ⊥ CDEF ⊥ CMF
TC ⊥ CD

Figure 40. Dissection to derive the dimensions of County B's contribution to the dyke, Problem 4.

interpreted it as fragments of a derivation of (2.77), but we have not been able to give a plausible guess as to how this derivation would proceed. One interesting aspect is that it begins with a description of a volume dissection, one of only two such descriptions in the *Jigu suanjing* (for the other see Sections 2.6.1.5 above and 3.3.3.4 below). See Figure 40:

> This 'dragon tail' resembles a *yanchu*. One *qiandu* and one *bienao* are put together side by side.

2.8. Problems 15–20: Right triangles

In each of Problems 15–20 (Sections 3.15–3.20) two relations between the sides of a right triangle are given, and the sides are required. In each case the calculation is derived by dissection of a three-dimensional object.[1]

As was mentioned in Section 1.4.2 above, in pre-modern Chinese mathematics the sides of a right triangle are referred to as *gou* 句 for the shorter leg, *gu* 股 for the longer leg, and *xian* 弦 for the hypotenuse. We translate these terms as 'base', 'leg', and 'hypotenuse', and in equations denote them as *a*, *b*, and *c* respectively; see Figure 41.

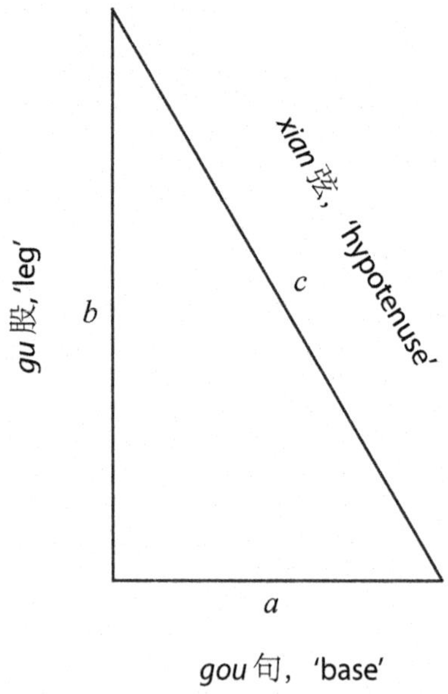

A peculiarity of the six triangle problems is that no units are given for the quantities involved. This is virtually unique in all of pre-modern Chinese mathematics (including Wang Xiaotong's Problems 1–14), in which problems are normally stated with reference to practical situations.

In **Problem 15** (Section 3.15) the given quantities are:

$$ab = 706 \tfrac{1}{50}$$
$$c-a = 36 \tfrac{9}{10}$$

and the values of *a*, *b*, and *c* are required. In the solution the coefficients of a cubic equation are calculated:

$$a^3 + \frac{c-a}{2}a^2 = \frac{(ab)^2}{2(c-a)} \tag{2.78}$$

$$a^3 + 18\tfrac{9}{20}a^2 = 6{,}754 \tfrac{129}{500}$$

This has one real root, $a = 14\tfrac{7}{20}$, and the remaining quantities are

$$b = \frac{ab}{a} = 49 \tfrac{1}{5}$$

$$c = a + (c-a) = 51 \tfrac{1}{4}$$

A comment in smaller characters (Section 3.15.1) explains (2.78) in a mixture of algebraic and geometric reasoning. It notes first that

$$(ab)^2 = a^2 b^2$$

gu 股, 'leg'

xian 弦, 'hypotenuse'

b

c

a

gou 句, 'base'

Figure 41. Terminology for right triangles, Problems 15–20.

---

1 On these problems see also Li Yan 1998: 137–138.

The comment then refers to a *fang* 方 (rectangular parallelepiped) and to quantities 'lined up'; this suggests a geometric rather than algebraic derivation of the method. We reconstruct this derivation as in Figure 42. We note that

$$\frac{a^2b^2}{2(c-a)} = \frac{a^2(c^2-a^2)}{2(c-a)} = a^2\frac{a+c}{2}$$

(which Wang Xiaotong does not state explicitly). Then using Figure 42,

$$\frac{(ab)^2}{2(c-a)} = a^3 + \left(\frac{c-a}{2}\right)a^2$$

It is not known whether Wang Xiaotong's book originally included illustrations like Figure 42. We think that it probably did not; he seems simply to give verbal descriptions of the geometric constructions which he uses.

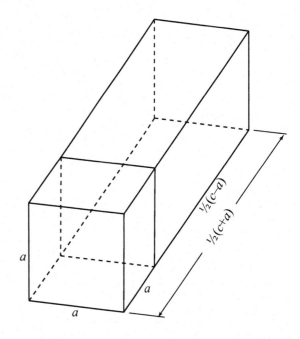

Figure 42. Construction to derive dimensions of right triangle, Problems 15–16.

Problem 16 (Section 3.16) is equivalent to Problem 15. The given quantities are

$$ab = 4{,}036\,^1/_5$$
$$c-b = 6\,^1/_5$$

and c is required. The cubic equation here is

$$b^3 + \frac{c-b}{2}b^2 = \frac{(ab)^2}{2(c-b)} \qquad (2.79)$$

$$b^3 + 3\,^1/_{10}\,b^2 = 1{,}313{,}783\,^1/_{10}$$

This has one real root, $b = 108\,^1/_2$, whence

$$c = b + (c-b) = 114\,^7/_{10}$$

Problem 17 (Section 3.17) gives the quantities

$$ac = 1{,}337\,^1/_{20}$$
$$c-b = 1\,^1/_{10}$$

and b is required. The cubic equation arrived at is

$$b^3 + \frac{5(c-b)}{2}b^2 + 2(c-b)^2 b = \frac{(ac)^2}{2(c-b)} - \frac{(c-b)^3}{2} \qquad (2.80)$$

$$b^3 + 2\,{}^3/_4 b^2 + 2\,{}^{21}/_{50} b = 812{,}591\,{}^{59}/_{125} \qquad (2.81)$$

This has one real root, $b = 92\,{}^2/_5$.

A comment explains (2.80) geometrically, but here our textual problems start, for a large number of characters are missing in the extant editions. In spite of the lacunae in the comment text it is clear that the geometrical construction is similar to that shown in Figure 43. The solid in Figure 43 has volume

$$\frac{c^2(c+b)}{2} - \frac{(c-b)^3}{2} = \frac{c^2(c^2-b^2)}{2(c-b)} - \frac{(c-b)^3}{2}$$

$$= \frac{a^2 c^2}{2(c-b)} - \frac{(c-b)^3}{2}$$

which is the right side of (2.80). The sum of the volumes of the blocks into which the solid is divided is

$$b^3 + 2(c-b)b^2 + \frac{c-b}{2}b^2 + (c-b)^2 b + 2\frac{(c-b)^2}{2}b$$

$$= b^3 + \frac{5(c-b)}{2}b^2 + 2(c-b)^2 b$$

which is the left side of (2.80).

Note that the dimensions of the triangles in Problems 15–17 are based on Pythagorean triples with c–b equal to 1 or 2:

15. $[14\,{}^7/_{20}, 49\,{}^1/_5, 51\,{}^1/_4] = [7, 24, 25] \times {}^{41}/_{20}$

16. $[37\,{}^1/_5, 108\,{}^1/_2, 114\,{}^7/_{10}] = [12, 35, 37] \times {}^{31}/_{10} \qquad (2.82)$

17. $[14\,{}^3/_{10}, 92\,{}^2/_5, 93\,{}^1/_2] = [13, 84, 85] \times {}^{11}/_{10}$

In **Problems 18–20** (Sections 3.18–3.20) so much is missing from the text that any reconstruction will to some extent be speculative. The Chinese mathematician Zhang Dunren 張敦仁 (1754–1834) and the Korean mathematician Nam Pyŏng-Gil 南秉吉 (1820–1869) have given reconstructions which agree in principle but differ in de-

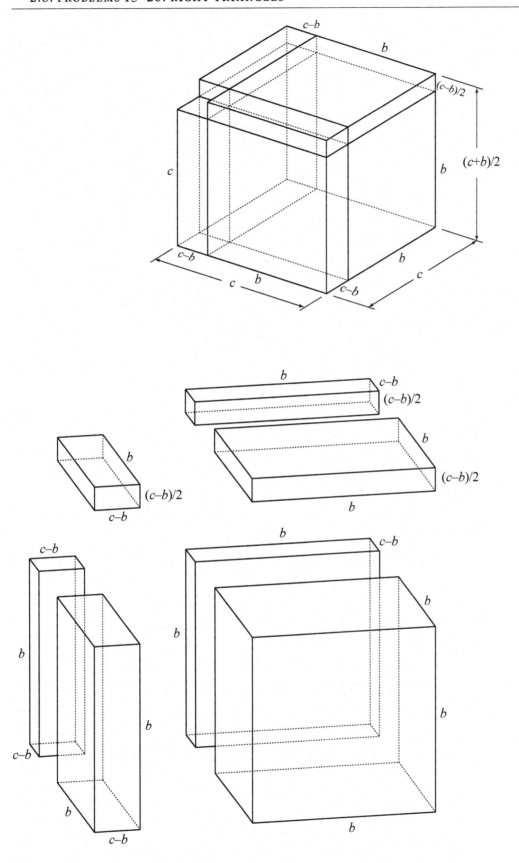

Figure 43. Construction to derive dimensions of right triangles, Problems 17–18.

tail. Neither explains how he arrived at his reconstruction, but some matters seem clear.

Any reconstruction should satisfy the approximate number of characters seen to be missing from the text. In addition, certain assumptions can safely be made: (1) The problems all concern right triangles; (2) the dimensions of the triangles are derived from Pythagorean triples, as in Problems 15–17; (3) the problem statements all follow the same rigid pattern as in Problems 15–17; and (4) the problems come in pairs, so that, just as 15 and 16 are equivalent, so are the pairs 17–18 and 19–20.

Assumption (4) implies that the fragmentary problem statements are

17. Given ac and $c-b$, determine b.
18. Given bc and $c-a$, [determine either a or b].
19. Given bc [and a], determine b.
20. Given [ac and] b, [determine a].

What remains of the 'method' for Problem 18 indicates that the quantity to be determined is b.

ZHANG DUNREN AND NAM PYŎNG-GIL agree on the form of the problem statements as given here, but they differ on the numerical values of the given quantities in Problems 18 and 19. Their reconstructions are shown in the table on the next page.

The Pythagorean triples on which these reconstructions are based are:

18. $[12\,^1/_{10}, 66, 67\,^1/_{10}] = [11, 60, 61] \times\,^{11}/_{10}$
$[15\,^3/_{10}, 68, 69\,^7/_{10}] = [9, 40, 41] \times\,^{17}/_{10}$
19. $[^7/_{100}, ^6/_{25}, ^1/_4] = [7, 24, 25] \times\,^1/_{100}$
$[7\,^7/_{10}, 26\,^2/_5, 27\,^1/_2] = [7, 24, 25] \times\,^{11}/_{10}$
20. $[8\,^4/_5, 16\,^1/_2, 18\,^7/_{10}] = [8, 15, 17] \times\,^{11}/_{10}$

In the following we have arbitrarily chosen to follow Zhang Dunren's reconstructions of the given quantities. Zhang Dunren also attempts, with much less success, to reconstruct large parts of the 'method' sections, and we have not followed him in the translation of these sections. Nam Pyŏng-Gil's edition does not include Wang Xiaotong's methods, but instead gives his own, which follow the methods of the traditional Chinese algebra of polynomials described in Section 2.10 below.

Problem 18 is, by assumption (4) above, equivalent to Problem 17. It results in the cubic equation

			Nam	Zhang
		Reconstructions of Problems 18–20 by Nam Pyŏng-Gil and Zhang Dunren		
18	Given	$bc =$	$4428\,{}^3/_5$	$4739\,{}^3/_5$
		$c{-}a =$	55	$54\,{}^2/_5$
	Result	$b =$	66	68
	Other quantities	$a =$	$12\,{}^1/_{10}$	$15\,{}^3/_{10}$
		$c =$	$67\,{}^1/_{10}$	$69\,{}^7/_{10}$
19	Given	$bc =$	${}^3/_{50}$	726
		$a =$	${}^7/_{100}$	$7\,{}^7/_{10}$
	Result	$b =$	${}^6/_{25}$	$26\,{}^2/_5$
	Other quantity	$c =$	${}^1/_4$	$27\,{}^1/_2$
20	Given	$ac =$	$164\,{}^{14}/_{25}$	
		$b =$	$16\,{}^1/_2$	
	Result	$a =$	$8\,{}^4/_5$	
	Other quantity	$c =$	$18\,{}^7/_{10}$	

$$a^3 + \frac{5(c-a)}{2}a^2 + 2(c-a)^2 a = \frac{(bc)^2}{2(c-a)} - \frac{(c-a)^3}{2}$$

(cf. (2.80) above). In Zhang Dunren's reconstruction,

$$a^3 + 136\,a^2 + 5{,}918\,{}^{18}/_{25}\,a = 125{,}974\,{}^{233}/_{1000}$$

This has one real root, $a = 15\,{}^3/_{10}$, and

$$b = \frac{bc}{(c-a)+a} = 68$$

Problems 19 and 20 are extremely fragmentary, but the assumption that they are equivalent makes Zhang Dunren's and Nam Pyŏng-Gil's reconstructions of the problems and their methods plausible. The method for Problem 19 leads in each case to the quadratic equation

$$x^2 + a^2 x = (bc)^2 \qquad (2.83)$$

in which $x = b^2$. In Zhang Dunren's version this gives

$$x^2 + 59\,{}^{29}/_{100}\, x = 52{,}706$$

This has one positive root, $x = 696\,{}^{24}/_{25}$, and $b = 26\,{}^2/_5$.

It is easy to derive (2.83) algebraically, for

$$b^4 + a^2 b^2 = b^2 \left(a^2 + b^2 \right) = (bc)^2$$

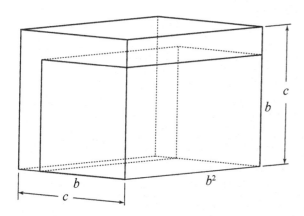

but what little remains of the comment implies a geometric explanation. Commentators have had difficulty here, for a geometric explanation would seem to involve a four-dimensional construction. The lack of units in the quantities involved, however, makes the construction shown in Figure 44 plausible. Here both b and b^2 are used as linear measures. The volume of the solid is

$$b^2 c^2 = (bc)^2$$

which is the right side of (2.83), and the sum of the two blocks into which the solid is divided is

$$b^4 + \left(c^2 - b^2 \right) b^2 = x^2 + a^2 x$$

Figure 44. Construction to derive dimensions of right triangles, Problems 19–20.

which is the left side of (2.83). If our reconstruction is correct, the 'length' mentioned in one of the remaining fragments of the comment is probably x, the length of the solid.

Problem 20 (Section 3.20) assumed to be equivalent to Problem 19, results in the quadratic equation

$$x^2 + b^2 x = (ac)^2$$

where $x = a^2$. In Zhang Dunren's reconstruction,

$$x^2 + 272\,{}^1/_4\, x = 27{,}079\,{}^{621}/_{625}$$

This has one positive root, $a = 8\,{}^4/_5$.

2.9. Pythagorean triples

We have noted in the previous section three disguised Pythagorean triples (right triangles with integral sides),

$$[7, 24, 25] \times {}^{41}/_{20}$$
$$[12, 35, 37] \times {}^{31}/_{10}$$
$$[13, 84, 85] \times {}^{11}/_{10}$$

Another shows up in Problems 3 and 5 (Section 2.6.1 above). See Figure 33. The triangle used in Problem 3, Equation (2.68) is

$$[l, h_2-h_1, s] = 10 \times [480, 31, 481]$$

and the corresponding triangles for the contributions of the counties, e.g. Equation (2.70), are

$$[l_A, h_A-h_1, s_A] = 4 \times [480, 31, 481]$$
$$[l_B, h_B-h_A, s_B] = 3 \times [480, 31, 481]$$
$$[l_C, h_C-h_B, s_C] = 2 \times [480, 31, 481]$$
$$[l_D, h_D-h_C, s_D] = 1 \times [480, 31, 481]$$

In the equivalent Problem 5, Figure 38, we find,

$$[l, h_2-h_1, s] = 7.2 \times [480, 31, 481]$$
$$[l_A, h_A-h_1, s_A] = 3 \times [480, 31, 481]$$
$$[l_B, h_B-h_A, s_B] = 2.4 \times [480, 31, 481]$$
$$[l_C, h_C-h_B, s_C] = 1.2 \times [480, 31, 481]$$
$$[l_D, h_D-h_C, s_D] = 0.6 \times [480, 31, 481]$$

A general method for generating Pythagorean triples seems to have been known to Chinese mathematicians since the *Jiuzhang suanshu*,[1] but these four triples, the ones known to have been used by Wang Xiaotong, all have $c-b$ equal to 1 or 2. They could therefore have been generated using the simpler special cases ascribed by Proclus to Pythagoras and Plato respectively:[2]

$$m^2 + \left(\frac{m^2-1}{2}\right)^2 = \left(\frac{m^2+1}{2}\right)^2$$

$$(2m)^2 + (m^2-1)^2 = (m^2+1)^2$$

---

1 Van der Waerden 1983: 5–7.

2 Heath 1921, 1: 80–81.

2.10. The traditional Chinese algebra of polynomials

In discussing Wang Xiaotong's 'reasoning about calculations' (Section 2.2.3 above) we mentioned and dismissed the question of whether this should be called algebra. An answer to that terminological question seems less interesting than attempting to understand, as precisely as we can, how Wang Xiaotong did his mathematics.

Let us nevertheless now consider this concept, *algebra*, and how the concept can be useful.[1] The classic discussion is Nesselmann's *Algebra der Griechen* (1842: 302), in which he defined three stages (*Stufen*) in the development of algebra: *rhetorical*, *syncopated*, and *symbolic* algebra. Wang Xiaotong's reasoning about calculations would seem to fit the definition of rhetorical algebra, but in China this was not a 'stage' in the development of the other two kinds of algebra. In fact, Wang Xiaotong appears to stand alone; there is very little else in pre-modern Chinese mathematical works which corresponds to any of Nesselmann's stages.[2]

It is more useful to expand the concept and define algebra as comprising all of the methods which have been used to solve problems which we in modern times solve using algebra.[3]

One method in the classic *Jiuzhang suanshu* which certainly belongs in this category is *fangcheng* 方程, 'rectangular arrays', in which systems of linear equations in several unknowns are laid out in a matrix on the counting board and solved by what amounts to Gaussian elimination.[4] Other methods in that book, for example *regula falsi*,[5] might also fall into this broad category.

It might be reasonably said that Wang Xiaotong's book presents approaches to a certain class of algebraic problems, always leading to a polynomial equation whose root can be approximated by one well-

---

1 In writing this section we have had the benefit of illuminating discussions with Jens Høyrup. Note also his *Algebra in cuneiform* (Høyrup 2013).

2 Some of Liu Hui's comments on the *Jiuzhang suanshu* may be seen as exceptions to this statement.

3 And of course we mean here 'elementary algebra' as it is taught in schools, not 'modern algebra', which 'treats classes of algebras having certain properties in common' (Pratt 2014).

4 Chapter 8, Qian Baocong 1963: 221–240; Chemla and Guo 2004: 599–659; Guo Shuchun et al. 2013: 904–932.

5 Chapter 7, *Yingbuzu* 盈不足, Qian Baocong 1963: 205–219; Chemla and Guo 2004: 549–597; Guo Shuchun et al. 2013: 800–903.

known method ('Horner's method', Section 1.5 above). Here it is important to note that Wang Xiaotong in his comments clearly shows a recognition that the arrangement of numbers on the counting board used in extracting roots of polynomials represents what we call an equation: it is in effect a statement that performing certain operations on an unknown quantity results in a certain known quantity.

Later, in works of the 13th century, we see an algebra of these polynomials, most often referred to as *tianyuan yi* 天元一.[1] Polynomials are added, subtracted, and multiplied in order to arrive at an 'equation' (arrangement of numbers on the counting board) whose roots can be approximated by the usual method. This algebra became quite sophisticated, and could involve polynomials in up to four variables.

In the 14th century the algebra of polynomials was almost totally forgotten, for reasons which we cannot begin to go into here. In the 18th and 19th centuries, after contact with Western mathematics, a number of Chinese mathematicians revived the old methods. One of these was Zhang Dunren, who, as noted in Section 1.6 above, published an edition of the *Jigu suanjing* in 1803. In his comments in this edition he shows how to solve each problem using the algebra of polynomials. This gives us a chance to see Wang Xiaotong's methods in the light of later developments.

HERE WE TAKE PROBLEM 17 as our example (Sections 2.8 above and 3.17 below). In modern terms the problem is, given the three equations

$$a^2 + b^2 = c^2$$
$$ac = 1337\,^1/_{20}$$
$$c - b = 1\,^1/_{10}$$

to find b. Wang Xiaotong converts this to a problem in solid geometry which he can solve by dissection (Figure 43, p. 105). He uses this approach in all of the triangle problems (15–20, Section 2.8 above) and elsewhere in the book, but he has no general method. Each problem requires the invention of an *ad hoc* method. Readers may wish, before reading further, to contemplate how they would solve Problem 17.

---

1 Martzloff 1997: 258–271; Chemla 1982; Guo Shuchun et al. 2006; Qian Baocong 1966b: 104–148, 166–209, 270–278; Mei Rongzhao 1966.

The algebra of polynomials developed in the Song and Yuan periods provides a general method for solving this sort of problem. Zhang Dunren in the 18th century learned this method and used it in solving all of Wang Xiaotong's problems. Though he had been exposed to some part of the Western mathematics of the time, there is nothing in his comment which could not have been written by a 13th-century Chinese mathematician.

The text is shown in Figure 45 and translated in Box 5. It can be seen that his method is quite general, and that he solves the problem in much the same way that a modern student might:

$$c = b + 1.1$$

$$c^2 = (b + 1.1)^2 = b^2 + 2.2b + 1.21$$

$$a^2 = c^2 - b^2 = 2.2b + 1.21$$

$$(ac)^2 = 2.2b^3 + 6.05b^2 + 5.234b + 1.461$$

$$= (1337.05)^2 = 1{,}787{,}702.7025$$

$$2.2b^3 + 6.05b^2 + 5.234b = 1{,}787{,}701.2384$$

After which b can be approximated using Horner's Method. A modern student would no doubt prefer to use the root-finding function of a pocket calculator.[1]

WANG XIAOTONG'S BOOK SHOWS US a stage in the development of this form of algebra. In his time it is recognized that the arrangement of numbers on the counting board used in extracting roots amounts to an 'equation'; a few centuries later, Chinese mathematicians will develop general methods of manipulating such equations to solve difficult algebraic problems.

---

1 A closed-form solution for the roots of a cubic exists (Weisstein 1996), but it is so complicated that it is rarely used except inside mathematical software.

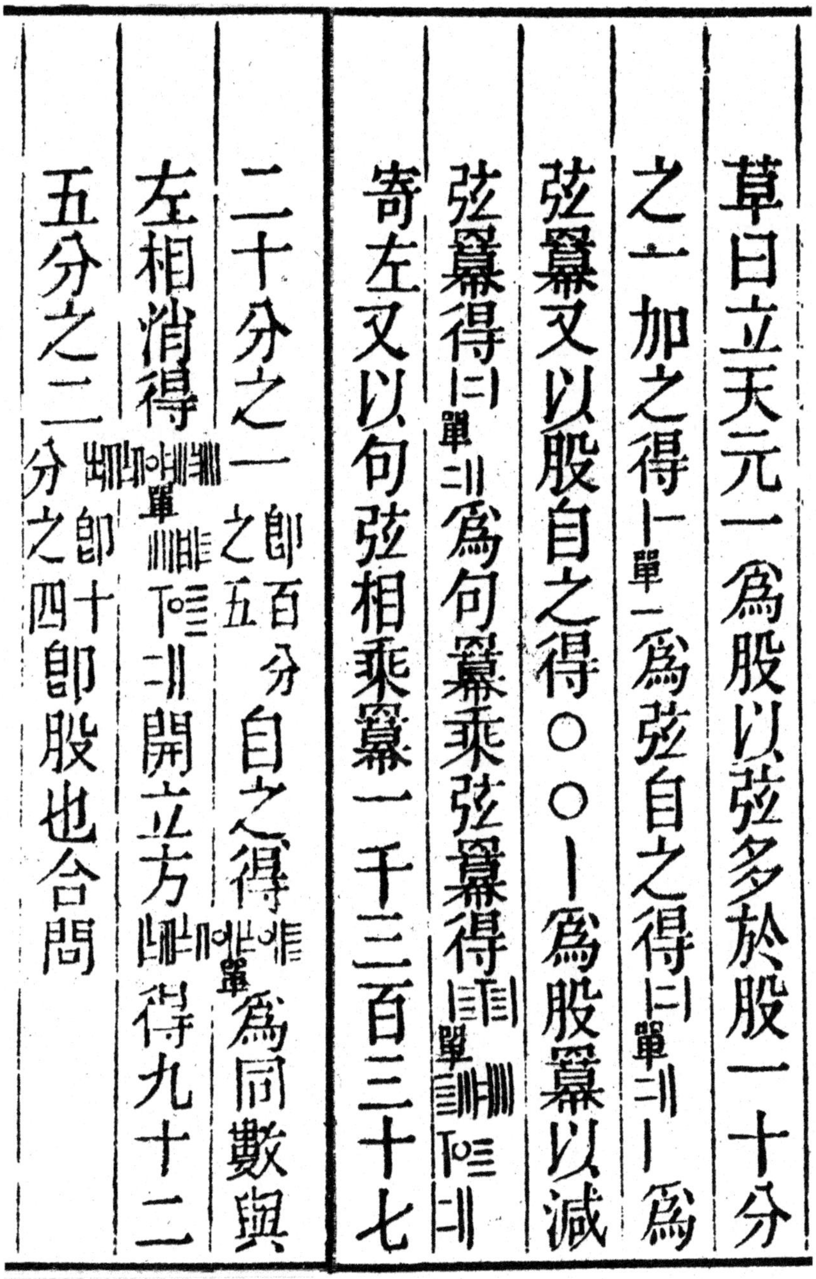

草曰立天元一爲股以弦多於股一十分
之一加之得一單一爲弦自之得一單二一爲
弦羃又以股自之得○○一爲股羃以減
弦羃得一單三爲句羃乘弦羃得
寄左又以句弦相乘羃一千三百三十七
二十分之一即百分自之得爲同數與
左相消得開立方得九十二
五分之二分之四即股也合問

Figure 45. Text of Zhang Dunren's algebraic derivation of the dimensions of a right triangle, Problem 17. From Zhang Dunren 1803: 128–129.

Box 5. Translation of Zhang Dunren's method for Problem 17

[The full text of Zhang Dunren's comment is shown in Figure 45; here can be seen the way he displays counting-board arrangements of numbers ('polynomials') as integral parts of his text. For the counting-rod notation see Box 1 on p. 22.]

Establish the *tianyuan yi* 天元一 [the unknown quantity to be solved for] as the leg [of the triangle, *b* in Figure 41, p. 102 above]. Adding to this the amount by which the hypotenuse is greater than the leg [*c–b*], $1^{1}/_{10}$, gives

	1.1	
單	1	$b + 1.1$

[The character *dan* 單, 'single', here indicates the units position in the counting-rod numbers.]

This is the hypotenuse [*c*]. Multiplying this by itself gives $[(b + 1.1)^2 =]$

	1.21	
單	2.2	$b^2 + 2.2b + 1.21$
	1	

This is the area of [the square on] the hypotenuse [c^2].
Multiplying the leg by itself gives

◯	0	
◯	0	$b^2 + 0b + 0$
│	1	

This is the area of [the square on] the leg [b^2]. Subtracting this from the area of [the square on] the hypotenuse gives [$c^2 - b^2 =$]

	1.21	
單	2.2	$2.2b + 1.21$

This is the area of [the square on] the base [a^2]. Multiplying by the area of [the square on] the hypotenuse [c^2] gives[1] [$a^2c^2 =$]

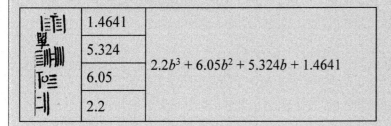	1.4641	
	5.324	$2.2b^3 + 6.05b^2 + 5.324b + 1.4641$
	6.05	
	2.2	

This is moved to the left.

The product of the base and the hypotenuse, $1,337\,^1/_{20}$ (i.e., $^5/_{100}$),[2] multiplied by itself, is[3] [$(ac)^2 =$]

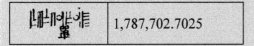	1,787,702.7025

This is the same number. Reducing it by what is on the left gives

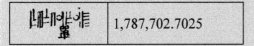	1,787,701.2384	
	5.324	$2.2b^3 + 6.05b^2 + 5.324b$
	6.05	$= 1,787,701.2384$
	2.2	

[This is equivalent to (2.81) on p. 104.]

Extracting the cube root gives $92\,^2/_5$ (i.e., $^4/_{10}$).[4] This is the leg [b], which is the answer.

---

1 Note scribal error in the counting-rod display, 6.04 for 6.05.

2 Comment in smaller characters.

3 Note scribal error in the counting-rod display, 1,787,202.7025 for 1,787,702.7025.

4 Comment in smaller characters.

Part III

Chinese text and translation

In the Chinese text we indicate variants using the form $\{a|b|c\}$, where *a* is the original as seen in the *Tianlu linlang* version (the version from which all other versions derive), *b* is Qian Baocong's (1963) version, and *c*, if present, is from another version, identified in a footnote. We copy Qian Baocong's punctuation throughout the Chinese text.

For the units of measure used see Box 3, p.36.

Mathematical comments by the translators are given indented in the text, while philological comments are given in footnotes.

Wang Xiaotong often names intermediate quantities in the calculation. In the translation we capitalize these names and denote them in equations with the symbols K_1, K_2, etc.

Commentators are divided on the function of the word *cong* (or perhaps *zong*) 從 in the recurring phrase 從開立方除. Qian Baocong (1966a: 46–47) reviews the various attempts to explain it, and comes to the conclusion that in each case it belongs at the end of the previous sentence, and means 'follow, accompany' (*gensui* 跟隨). This is difficult for us to understand, and we omit the word and translate the phrase as 'Extract the cube root.'

WE HAVE DIVIDED THE TRANSLATION into numbered sections in order to ease understanding of Wang Xiaotong's intentions. The titles of these sections are not part of the original text.

3.0. Introduction: Wang Xiaotong's presentation of his book to the Throne

The translation here is freer than in the strictly mathematical parts of the text.

3.0.1. Initial obeisances

Your servant Xiaotong states: I have read in works on the art of government that the law is manifested in the cardinal human relations. In the completion of the Six Arts, the Art of Numbers participates in the transformation.

In the activities of a ruler, regulating the people, having the divine Way, establishing his teaching, appointing competent and talented [ministers], fulfilling his nature, and exhaustively seeking out the origins [of matters and things], nothing is more important than mathematics.

3.0.2. The Nine Chapters, Jiuzhang suanshu 九章算術

Of old, in the Rites established by the Duke of Zhou, are mentioned the 'Nine Numbers'.[1] My humble investigations indicate that 'the Nine Numbers' refers in fact to the *Nine Chapters* [i.e. the *Jiuzhang suanshu*].[2] [This book's] principles are profound and precise, [but] its form is obscure and abbreviated.

Chonggou 重句 [similar triangles?] can be used to survey the ocean,[3] and with a wooden stick the heavens can be measured. If this is not the most precise in the universe, who can improve it?

In the Han period Zhang Cang[4] 張蒼 filled out its lacunae and compared its individual entries, so that its methods are considerably different from the ancient ones.

上緝古算經表

臣孝通言：臣聞九疇載敍，紀法著於彝倫；六藝成功，數術參於造化。夫為君上者司牧黔首，{布∣有}神道而設教，采能事而經綸，盡性窮源莫重於算。

昔周公制禮，有九數之名。竊尋九數即「九章」是也。其{禮∣理}幽而微，其形{祕∣秘}而約，重句聊用測海，寸木可以量天，非宇宙之至精，其孰能與於此者。漢代張蒼刪補殘缺，校其條目，頗與古術不同。魏朝劉徽篤好斯言，{博∣博}綜纖隱，更為之注。徽思極毫芒，觸類增長，乃造重差之法，列於終篇。雖即未為司南，然亦一時獨步。自茲厥後，不繼前蹤。賀循、徐岳之徒，王彪、甄鸞之輩，會通之數無聞焉耳。但舊經殘駮，尚有闕漏。自劉{已∣以}下更不足言。

---

1 A reference to 'The rites of Zhou', *Zhou li* 周禮, j. 14, *Shisanjing zhushu* 1980: 731b. The teaching of the sons of the elite includes the 'nine numbers', *jiu shu* 九數. Translation Biot 1851: 296–298.

2 On this book see Section 1.4 above.

3 As in Liu Hui's *Haidao suanjing*, see fn. 2 on p. 120 below.

4 Zhang Cang was an important official at the beginning of the Han dynasty, among other things involved in the reform of the calendar (Loewe 2000: 675–676). Liu Hui's preface to his edition of the *Jiuzhang suanshu* names

Under the Wei Dynasty, Liu Hui[1] 劉徽 highly esteemed this text. He arranged its subtleties and produced a new commentary. His reasoning was extremely meticulous, and he also enlarged the subject matter, adding the 'Method of multiple differences' [*Chongcha zhi fa* 重差之法] as a final chapter.[2] Though this was still not a perfect guide, it was unrivalled for a time.

From this time on the ancient traces were not followed up. Among the followers of He Xun 賀循 and Xu Yue 徐岳, and the circle of Wang Biao 王彪 and Zhen Luan 甄鸞, none has been heard of who thoroughly mastered mathematics.[3] But the ancient classic is fragmentary and contradictory, and there are still important gaps; after Liu Hui's time this has become even worse.

3.0.3. *Zhui shu* 綴術

其祖暅之「綴術」，時人稱之精妙。曾不覺方邑進行之術全錯不通，芻亭方亭之問於理未盡。臣今更作新術，於此附伸。

The book *Zhui shu* by Zu Xuan[4] 祖暅 was praised in its time as wonderfully precise, but it was not realized that the method of 'entering a square city' was completely incorrect and illogical, and that the problems concerning *chuting* 芻亭 and *fangting* 方亭 do not exhaust the principles.[5]

Your servant has produced new methods, and [my book] is attached herewith.

3.0.4. *Curriculum vitae*

臣長自閭閻，少小學筭，鐫磨愚鈍，迄將{皓|皎}首，鑽尋祕{奧|奧}，曲盡無遺。代乏知音，終成寡和。伏蒙聖朝收拾，用臣為太史丞。比年已來，奉{敕|勅}校勘傅仁均曆，凡駁正術錯三十餘道，即付太史施行。

Your servant grew up among the common people. From an early age I have studied calculation, and have exercised my slow wits

him as one of its early editors (Qian Baocong 1963: 91; Chemla and Guo 2004: 43–56, 127; Guo et al. 2013: 9).

1   See Section 1.4 above.

2   Liu Hui's additional chapter was later removed from the *Jiuzhang suanshu* and published separately as *Haidao suanjing* 海島算經, 'Sea island computation canon'.

3   Very little is known about these persons. He Xun was an official dealing with calendrical matters in the Liu-Song dynasty (420–479). Xu Yue was the author of *Shushu jiyi* (see Section 1.3 above). Wang Biao was a military official in the Liu-Song and Southern Qi dynasties (420–479, 479–502). Zhen Luan wrote commentaries on several important mathematical textbooks. See Li Di 1999: 199 and Martzloff 1997: 124–125.

4   This book is usually attributed to Zu Xuan's father, Zu Chongzhi 祖冲之. It has long been lost, and its subject, and even the translation of the title, are matters of controversy. See Section 1.3.1 above.

5   See Box 4, p. 38. *Chuting* is probably a scribal error for *chutong* 芻童, or possibly *chumeng* 芻甍 (Wang Xiaoqin 1999).

until now, when my hair is turning white. I have sought out its profundities exhaustively, until there is no remainder. This age lacks comprehension, and therefore few understand [my work].

Your servant humbly received the Imperial Court's summons, and was appointed to the post of Assistant to the Grand Astrologer. In recent years I have had the task of verifying the calendrical system of Fu Renjun 傅仁均; I corrected over thirty errors in method, and presented these to the Grand Astrologer for implementation.[1]

3.0.5. *This book*

I have humbly investigated the chapter 'Calculating labour quotas' of the *Jiuzhang* [*suanshu*]. Here there is a method for calculating the length on level ground [of a construction] assigned to [labourers owing] a quota of labour. But if [the construction in question] is wide above and narrow below, or high at the front and low at the back, the book lacks [a method] and does not consider it.[2] As a result, modern persons, misunderstanding the deeper principles, treat crooked paths as if they were straight. This is 'putting a square peg into a round socket'; how could it be acceptable?

Your servant has cogitated day and night. I look at what I have written and sigh, fearing that one day I will 'close my eyes', and that in the future no one will see it. Therefore after [dealing with] the addition to [methods for] level ground, I have continued with methods for narrow and slanted [constructions].

In all there are twenty methods, and the title of the book is *Jigu*, 'Continuing the ancient'. I beg that persons may be consulted who are skilled in calculation, to investigate its merits and demerits. If one of them can fault one word, your servant wishes to reward him with a thousand *jin* 金.[3]

Requesting[4] that what I have presented be employed, I prostrate myself and shake with fear.

Respectfully submitted.

伏尋「九章」商功篇有平地役功受袤之術。至於上寬下狹、前高後卑,正經之內闕而不論。致使今代之人不達深理,就平正之間同敧邪之用。斯乃圓孔方柄,如何可安?臣晝思夜想,臨書浩歎,恐一旦瞑目,將來莫覩。遂於平地之餘,續狹斜之法,凡二十術,名曰「緝古」。請訪能籌之人考論得失,如有排其一字,臣欲謝以千金。輕用陳聞,伏深戰悚。謹言。

---

1 See Section 1.1 above.

2 See Sections 1.2.1 and 1.2.2 above.

3 A thousand taels of silver, an enormous sum. (Taken literally, this means ca 41 kilograms of silver.)

4 Reading 請 for 輕.

緝古算經
唐通直郎太史丞臣王孝通撰并注

Continuation of ancient mathematics
Written and commented by Wang Xiaotong, Court Gentleman for Comprehensive Duty[1] and Assistant to the Grand Astrologer in the Tang period.

This heading was presumably added by the Northern Song publishers.

3.1. Problem 1: The sun's motion on the ecliptic

This problem does not concern geometry or polynomial equations, and is thus very different from the rest of the book; it is translated here only for the sake of completeness, and we have not attempted to understand completely the astronomical terminology and constants used, which appear to be non-standard. We are in fact in doubt as to whether this problem was a part of Wang Xiaotong's original text.

Andrea Bréard (1999: 333–336; 2002: 61, 73) gives a partial translation and analysis which have been useful to us.

3.1.1. The problem

假今天正十一月朔、夜半，日在斗十度、七百分度之四百八十。以章歲為母，朔月行定分九千，朔日定小餘一萬，日法二萬，章歲七百，亦名行分{也|法}。今不取加時{|日}度。問天正朔夜半之時，月在何處？

At the beginning of the year, the first day of the 11th month, at midnight, the sun's location [on the ecliptic] is at [$a =$] $10^{480}/_{700}$ degrees [*du* 度] in [the lunar mansion] *Dou* 斗.

Du is a unit of angular measurement: a circle is $365^{1}/_{4}$ *du*, so that one *du* ≈ 0.99 degree, and it is therefore convenient to translate the word as 'degree'. The ecliptic was divided into 28 *xiu* 宿 (usually translated 'lunar mansions'), of which *dou* was one.

If the 'year cycle' [*zhangsui* 章歲] [*y*] is taken as the denominator, then the 'determined numerator of the moon's motion on the

---

1 See Hucker 1985: 554.

first day' [*shuo yue xing ding fen* 朔月行定分] is [*m* =] 9,000, the 'determined small remainder on the first day' [*shuo ri ding xiao yu* 朔日定小餘] is [*r* =] 10,000, and the 'solar divisor' [*ri fa* 日法] is [*s* =] 20,000. The 'year cycle' is [*y* =] 700; it is also the denominator of the numerators of motion [*ay*, *by*, and *m*].[1] Without considering the addition of 'seasonal solar degrees' [*bu qu jia shi ri du* 不取加時日度], what is the location of the moon at midnight on the first day of the year?

3.1.2. Suspect comment in smaller characters

We agree with Qian Baocong (1963: 495) that the immediately following comment, supposedly by Wang Xiaotong, is 'confused'. It is unlikely to have been written by the author of the main text. We give here a fairly close translation (distinguished from other comment translations by a sideline) of what appears to be nonsense.[2]

For the calculation of the degrees of the moon at midnight of the first day of the month, the older method required the addition of 'seasonal solar degrees'. Since ancient times, though the earlier savants have repeatedly revised and changed the system, there have been numerous proposals, but none has been able to calculate it precisely. Their principles are not exhaustive, and it is difficult to determine what is correct. Your servant has cogitated day and night, and has constantly feared that the theories are so abstruse that in later generations there will be no one who understands them.

Having been ordered by Your Majesty to create a new calendar, I have now made a revision and created this new method. In the older method of calculating the degrees of the sun, after the 'degrees of the sun at midnight of the first day' has been obtained, it is necessary to add the 'seasonal solar degrees'; only then can the position of the moon be known. Your servant has now created a new method: it is only

推朔夜半月度，舊術要須加時日度。自古先儒雖復修撰改制，意見甚眾，竝未得算妙，有理不盡，考校尤難。臣每日夜思量，常以此理屈滯，恐後代無人知者。今奉勅造曆，因即改制，為此新術。舊推日度之術，已得朔夜半日度，仍須更求加時日度，然知月處。臣今作新術，但得朔夜半日度，不須加時日度，即知月處。此新術比於舊術，一年之中十二倍省功，使學者易知。

---

1 *Yi ming xing fen ye* 亦名行分也. Qian Baocong, Guo Shuchun, and Andrea Bréard follow Li Huang in emending *ye* 也 to *fa* 法, and Bréard (1999: 334) translates, 'Ihr Name ist auch "Divisor der gewanderten Teile"'. We have not accepted this emendation; we take *ming* 名 here to be a transitive verb, equivalent to *ming* 命, 'to denominate, be the denominator of (a fraction)'.

2 The general problem of 'ignorant editors', attempting to make sense of mathematical texts which they fail to understand, has been discussed by Wagner (2012a; 2012b).

necessary to obtain the 'degrees of the sun at midnight of the first day', without adding the 'seasonal solar degrees', and the position of the moon is known. Comparing this new method with the ancient methods, in the course of a year the saving in labour is twelvefold, and those who study it will find it easy to understand.

3.1.3. Answer

答曰：在斗四度、七百分度之五百三十。

[The moon is] in *Dou*, [*b* =] $4\,{}^{530}/_{700}$ degrees.

This answer is consistent with Equation (3.2) below, $b = a – f/y$, where f is defined in (3.1).

As Andrea Bréard (2002) points out, this is one of many 'pursuit problems' to be found in both Chinese and European mathematics. The sun starts at location a at midnight and moves along the ecliptic. The moon moves in the same direction at a speed which is m/y times that of the sun, and the two meet at a certain time (the exact time of the new moon, potentially a solar eclipse). The question is then the moon's location at midnight, the start of the 'pursuit'. The time at which the sun and the moon meet is obscure in the text, but reckoning backward from the given answer and method indicates that they meet at time $r/s = {}^1/_2$ day after midnight.

In the method described below, working with fractions is avoided until the end by letting some quantities be represented as improper fractions with $y = 700$ as their common denominator (ay/y, by/y, m/y, and [below] f/y).

3.1.4. Method

◆ 3.1.4.1. Comment in smaller characters

術曰：

推朔夜半月度新術不{復|須}加時日度，{月蝕|有定小餘}乃可用之。

In calculating the 'degrees of the moon at midnight of the first day' it is not necessary to add the 'seasonal solar degrees'. If one has the 'determined small remainder' [*r*], it can be used [instead].

以章歲減朔月行定分，餘以乘朔日定小餘，滿日法而一，為先行分。不盡者，半法已上收成一，已{上|下}者棄之。若先行分滿日行分而一，為度分，以減朔日夜半日所在度分。若度分不足減，加往宿度。其分不足減者，退一度為行分而減之。餘即朔日夜半月行所在度及分也。

◆ 3.1.4.2. Method

Subtract the 'year cycle' [*y*] from the 'determined numerator of the moon's motion on the first day' [*m*] and multiply the difference by the 'determined small remainder on the first day' [r]. Divide by the 'solar divisor' [s] to make the 'numerator of motion in advance' [xian xing fen 先行分] [f]. If the remainder is greater than half of the divisor, add one; if less, discard it.

$$f = \left(m - y \right) \frac{r}{s}, \text{ rounded up or down} \qquad (3.1)$$

If the 'numerator of motion in advance' [f] is less than the 'numerator of solar motion' [ri xing fen 日行分] [ay, the numerator of the improper fraction $a = 7{,}480/700 = 10^{480}/_{700}$ degrees], then divide by [the 'year cycle', y] to make [a number of] degrees with a fraction. Subtract this from the degrees with fraction of the location of the sun at midnight of the first day [a]. If the degrees with fraction cannot be subtracted, add the number of degrees of the previously-passed lunar mansion [l]. If the fractional part [of the subtrahend] is insufficient to be subtracted [from the fractional part of the minuend], convert one degree to a fraction[, add to the fractional part of the minuend,] and subtract. The difference is [b=] the location of the moon in its path in degrees with a fraction.

$$b = \begin{cases} a - \dfrac{f}{y} & , \text{if } \dfrac{f}{y} \le a \\[3mm] a + l - \dfrac{f}{y} & , \text{if } \dfrac{f}{y} > a \end{cases} \qquad (3.2)$$

In the case $f/y > a$, the moon at midnight was in the previously passed lunar mansion, whose width in degrees along the ecliptic is l.

3.1.5. Comment in smaller characters

In calculating the calendar, the 'determined numerator of the moon's motion' [m] is the numerator of the motion of the moon in one day. This 'determined numerator' alone, divided by the 'year cycle' [y], gives the number of degrees. The sun moves one degree in one day, and therefore the 'year cycle' [y] is [also] the 'numerator of the sun's motion in one day'.

The chapter 'Equable transportation' [Junshu 均輸] of the Nine chapters [Chapter 6 of Jiuzhang suanshu 九章算術] gives a method for 'a hound pursuing a hare' which is similar to this method.

凡入曆當月行定分即是月一日之行分。但此定分滿章歲而一，為度。凡日一日行一度。然則章歲者即是日之一日行分也。今按《九章·均輸篇》有犬追兔術，與此術相似。彼問：犬走一百走，兔走七十步。今兔先走七十五步，犬始追之，問幾何步追及？答曰：二百五十步追及。彼術曰：以兔走減犬走，餘者為法。又以犬走乘兔先走為實。實如法而一，即得追及步數。

The extant text of chapter 6 of the *Jiuzhang suanshu* includes a problem concerning 'a hound pursuing a hare' (Problem 14, Qian Baocong 1963: 192–193; Guo and Chemla 2004: 521; Guo Shuchun et al. 2013: 727–729; Bréard 1999: 337; discussion Bréard 2002: 60–62), but not the problem described in the following (see Guo and Chemla 2004: 842–843 n. 139).

There the question is: When the hound runs [$p =$] 100 paces, the hare runs [$q =$] 70 paces. Suppose the hare has run [$u =$] 75 paces when the hound begins to pursue it. After how many paces does it reach [the hare]? Answer: It is reached after 250 paces. The method there is: Subtract the [distance] run by the hare [q] from that run by the hound [p] and let the difference be the divisor. Further multiply the [distance run by the hound [p] by the [distance] which the hare ran in advance [u] to make the dividend. Divide the dividend by the divisor; the result [$pu/(q-p) = 250$] is the number of paces to reach [the hare].

3.1.6. Suspect comment in smaller characters

The sidelined comment that follows here seems again to be suspect. It is difficult to reconcile it with the main text.

此術亦然。何者？假令月行定分九千，章歲七百，即是日行七百分，月行九千分。今日月行數相減，餘八千三百分者，是日先行之數。然月始追之必用一日而相及也。{令|今}定小餘者，亦是日月相及之日分。假令定小餘一萬，即相及定分，此乃無對為數。其日法者，亦是相及之分，此又同數，為有八千三百是先行分也。斯則異矣。但用日法除之，{即|得}四千一百五十，即先行分。故以夜半之時日在月前，月在日後，以日月相去之數四千一百五十，減日行所在度分，即月夜半所在度分也。

The method given here is similar. Why is this? If the 'determined numerator of the moon's motion' [m] is 9,000, and the 'year cycle' [y] is 700, then this is [the same as saying] that when the sun moves 700 the moon moves 9,000. [Thus m is analogous to p and y is analogous to q.] Subtracting the values for the motion of the sun and moon gives a difference of [$m - y =$] 8,300 parts, and this is the value of the sun's motion in advance. However [? *ran* 然], for the start of the moon's 'pursuit' it is necessary to assume that it reaches [the sun] in one day. In the present case the 'determined small remainder' [r] is the numerator of the [number of] days to the meeting of the sun and moon.

If the 'determined small remainder' [r] is 10,000, and this is the 'determined numerator' of their meeting, this is a value without a corresponding [denominator]. The 'solar divisor' [s] is also the parts [*fen* 分, in this case the denominator?] of the meeting, and this is again an analogous number [? *tong shu* 同數], since [? *wèi* 為] 8,300 [$= m - y$] is the 'numerator of motion in advance'.

In this case it is different. Simply dividing by the 'solar divisor' [s] gives 4,150 [= $r(m-y)/s$], which is the 'numerator of motion in advance' [f]. Therefore, since at midnight the sun is ahead of the moon and the moon is behind the sun, subtracting the value of the distance between the sun and the moon, [f =] 4,150, from the numerator of the [number of] degrees of the sun's position [ay] gives the numerator of the [number of] degrees of the moon's position at midnight [by].

3.2. Problem 2: The Grand Astrologer's platform and ramp

3.2.1. The problem

◆ *3.2.1.1. The observatory platform*

In the following, consult Section 2.5.1 above and Figures 23 and 24, pp. 64 and 65.

假令太史造仰觀臺，上廣表少，下廣表多。上下廣差二丈，上下表差四丈，上廣表差三丈，高多上廣一十一丈。甲縣差一千四百一十八人，乙縣差三千二百二十二人，夏程人功常積七十五尺，限五日役臺畢。

Suppose that the Grand Astrologer builds a platform for observing the heavens. The upper width and length are smaller, while the lower width and length are larger. The difference between the upper and lower widths is [$a_2-a_1 =$] 2 *zhang* [= 20 *chi*], the difference between the upper and lower lengths is [$b_2-b_1 =$] 4 *zhang* [= 40 *chi*], the difference between the upper width and length is [$b_1-a_1 =$] 3 *zhang* [= 30 *chi*], and the height exceeds the upper width by [$h-a_1 =$] 11 *zhang* [= 110 *chi*].

County A sends 1,418 corvée labourers and county B sends 3,222 corvée labourers. The norm for one man's labour in summer is a constant volume of 75 [cubic] *chi* [of earth per day]. After 5 days the work on the platform is finished.

◆ *3.2.1.2. The ramp*

羨道從臺南面起，上廣多下廣一丈二尺，少表一百四尺，高多表四丈。甲縣一十三鄉，乙縣四十三鄉，每鄉別均賦常積六千三百尺，限一日役羨道畢。二縣差到人共造仰觀臺，二縣鄉人共造羨道，皆從先給甲縣，以次與乙縣。臺自下基給高，道自初登給表。問臺道廣、高、表及縣別給高、廣、表各幾何？

A ramp rises from the southern surface of the platform. The upper width is [$c_1-c_2 =$] 1 *zhang* 2 *chi* [= 12 *chi*] longer than the lower width, and [$l-c_1 =$] 104 *chi* shorter than the length [*l*]. The height exceeds the length by [$g-l =$] 4 *zhang* [= 40 *chi*].

County A [sends labourers from] 13 districts and county B [sends labourers from] 43 districts. Each district is uniformly assessed with a constant volume of 6,300 [cubic] *chi* [of earth]. After one day the work on the ramp is finished. The delegate corvée labourers from the two counties work together to build the platform. The men from the two counties' districts work together to build the ramp. In both cases the work is first assigned to county A and then to county B. The platform is built from the base upward, while the ramp is built from the beginning of elevation along the length.[1] What are the width, height and length

---

1 *Tai zi xia ji ji gao, dao zi chu deng ji mao* 臺自下基給高，道自初登給表. The word *ji* 給 would seem to have some specific meaning in the administrative terminology of the time.

of the platform and the ramp, and how much does each of the counties contribute in height, width and length?

3.2.2. Answer

The platform:

> height [$h =$] 18 *zhang* [= 180 *chi*];
> upper width [$a_1 =$] 7 *zhang* [= 70 *chi*];
> lower width [$a_2 =$] 9 *zhang* [= 90 *chi*];
> upper length [$b_1 =$] 10 *zhang* [= 100 *chi*];
> lower length [$b_2 =$] 14 *zhang* [= 140 *chi*].

The contribution of county A:

> height [$h_A =$] 4 *zhang* 5 *chi* [= 45 *chi*];
> upper width [$a_3 =$] 8 *zhang* 5 *chi* [= 85 *chi*];
> lower width [$a_2 =$] 9 *zhang* [= 90 *chi*];
> upper length [$b_3 =$] 13 *zhang* [= 130 *chi*];
> lower length [$b_2 =$] 14 *zhang* [= 140 *chi*].

The contribution of county B:

> height [$h_B =$] 13 *zhang* 5 *chi* [= 135 *chi*];
> upper width [$a_1 =$] 7 *zhang* [= 70 *chi*];
> lower width [$a_3 =$] 8 *zhang* 5 *chi* [= 85 *chi*];
> upper length [$b_1 =$] 10 *zhang* [= 100 *chi*];
> lower length [$b_3 =$] 13 *zhang* [= 130 *chi*].

The ramp:

> height [$g =$] 18 *zhang* [= 180 *chi*];
> upper width [$c_1 =$] 3 *zhang* 6 *chi* [= 36 *chi*];
> lower width [$c_2 =$] 2 *zhang* 4 *chi* [= 24 *chi*];
> length [$l =$] 14 *zhang* [= 140 *chi*].

The contribution of the men in the district of county A:

> height [$g_A =$] 9 *zhang* [= 90 *chi*];
> upper width [$c_3 =$] 3 *zhang* [= 30 *chi*];
> lower width [$c_2 =$] 2 *zhang* 4 *chi* [= 24 *chi*];
> length [$l_A =$] 7 *zhang* [= 70 *chi*].

The contribution of the men in the district of county B:

> height [$g_B =$] 9 *zhang* [= 90 *chi*];
> upper width [$c_1 =$] 3 *zhang* 6 *chi* [= 36 *chi*];
> lower width [$c_3 =$] 3 *zhang* [= 30 *chi*];
> length [$l_B =$] 7 *zhang* [= 70 *chi*].

答曰：
　臺高一十八丈，
　　上廣七丈，
　　下廣九丈，
　　上袤一十丈，下袤一十四丈。
　甲縣給高四丈五尺，
　　上廣八丈五尺，
　　下廣九丈，
　　上袤一十三丈，
　　下袤一十四丈。
　乙縣給高一十三丈五尺，
　　上廣七丈，
　　下廣八丈五尺，
　　上袤一十丈，
　　下袤一十三丈。
　羨道高一十八丈，
　　上廣三丈六尺，
　　下廣二丈四尺，
　　袤一十四丈。
　甲縣鄉人給高九丈，
　　上廣三丈，
　　下廣二丈四尺，
　　{上|}袤七丈{，下袤一十四丈|}。
　乙縣鄉人給高九丈，
　　上廣三丈六尺，
　　下廣三丈，
　　{下|}袤七丈。

3.2.3. Method

◆ *3.2.3.1. The dimensions of the platform*

術曰：以程功尺數乘二縣人，又以限日乘之，為臺積。又以上下表差乘上下廣差，三而一為隅陽冪。以乘截高，為隅陽截積｛冪｝。又半上下廣差，乘斬上表為隅頭冪，以乘截高為隅頭截積。｛所得｝并二積，以減臺積，餘為實。以上下廣差并上下表差，半之為正數。加截上表，以乘截高，所得，增隅陽冪加隅頭冪，為方法。又并截高及截上表與正數，為廉法，從。開立方除之，即得上廣。各加差，得臺下廣及上下表、高。

Multiply the labour norm in [cubic] *chi* by the number of men of the two counties. Multiply this by the allotted number of days to make the volume [V] of the platform.

$$V = 75 \; chi^3/\text{man·day} \times 4{,}640 \text{ men} \times 5 \text{ days}$$
$$= 1{,}740{,}000 \; chi^3$$

Multiply the difference between the upper and lower lengths [$b_2 - b_1$] by the difference between the upper and lower widths [$a_2 - a_1$] and divide by three to make [$K_1 =$] the Area for the Corner *Yang*[*ma*].[1] Multiply this by the truncated height[2] [$h - a_1$] to make [$K_2 =$] the Truncated Volume for the Corner *Yang*[*ma*].

$$K_1 = \frac{(a_2 - a_1)(b_2 - b_1)}{3} = 266^2/_3 \, chi^2$$

$$K_2 = K_1(h - a_1) = 29{,}333^1/_3 \, chi^3$$

Halve the difference between the upper and lower widths [$a_2 - a_1$] and multiply by the shortened upper length [$b_1 - a_1$] to make [$K_3 =$] the Area for the Corner End. Multiply by the truncated height [$h - a_1$] to obtain [$K_4 =$] the Truncated Volume for the Corner End.

$$K_3 = \frac{(a_2 - a_1)(b_1 - a_1)}{2} = 300 \, chi^2$$

$$K_4 = K_3(h - a_1) = 33{,}000 \, chi^3$$

Add the two volumes [K_2 and K_4] and subtract from the volume of the platform [V]. The remainder is the *shi* 實 [the constant term of the cubic equation].

$$shi = V - (K_2 + K_4) = 1{,}677{,}666^2/_3 \, chi^3$$

---

1 In keeping with the writing fashion of his time, in which polysyllabic words were considered inelegant, Wang Xiaotong throughout the book abbreviates the terms *yangma* and *bienao* to *yang* and *bie* respectively.

2 *Jiegao* 截高. The same name is given to a different quantity in Section 3.2.3.3 below. See fn. 1, p. 135.

Add the difference between the upper and lower widths $[a_2-a_1]$ to the difference between the upper and lower lengths $[b_2-b_1]$ and halve [the sum] to obtain $[K_5 =]$ the Determined Number.

$$K_5 = \frac{(a_2 - a_1) + (b_2 - b_1)}{2} = 30 \; chi$$

Add the truncated upper length $[b_1-a_1]$ and multiply by the truncated height $[h-a_1]$. To the result of this, add the sum of the Area for the Corner *Yang[ma]* $[K_1]$ and the Area for the Corner End $[K_3]$ to make the *fangfa* 方法 [the coefficient of the linear term].

$$fangfa = \left[K_5 + (b_1 - a_1)\right](h - a_1) + (K_1 + K_3)$$
$$= 7{,}166 \, {}^2/_3 \; chi^2$$

Further add together the truncated height $[h-a_1]$, the truncated upper length $[b_1-a_1]$, and the Determined Number $[K_5]$ to make the *lianfa* 廉法 [the coefficient of the quadratic term].

$$lianfa = (h-a_1) + (b_1-a_1) + K_5 = 170 \; chi$$

Extract the cube root to obtain the upper width $[a_1]$.

$$a_1^3 + \left[(h-a_1) + (b_1-a_1) + K_5\right]a_1^2$$
$$+ \left\{\left[K_5 + (b_1-a_1)\right](h-a_1) + K_1 + K_3\right\}a_1$$
$$= V - (K_2 + K_4)$$

$$a_1^3 + 170a_1^2 + 7{,}166⅔ \, a_1 = 1{,}677{,}666⅔ \; chi^3$$

$$a_1 = 70 \; chi$$

Add to each of the differences to obtain the lower width, the upper and lower lengths, and the height of the platform.

$$a_2 = (a_2-a_1) + a_1 = 90 \; chi$$
$$b_1 = (b_1-a_1) + a_1 = 100 \; chi$$
$$b_2 = (b_2-b_1) + b_1 = 140 \; chi$$
$$h = (h-a_1) + a_1 = 180 \; chi$$

求均給積尺受廣表術曰：以程功尺數乘乙
縣人，又以限日乘之，為乙積。三因之，
又以高冪乘之，以上下廣差乘表差而一，
為實。又以臺高乘上廣，廣差而一，為上
廣之高。又以臺高乘上表，表差而一，為
上表之高。又以上廣之高乘上表之高，三
之，為方法。又并兩高，三之，二而一，
為廉法，從。開立方除之，即乙高。以減
本高，餘，即甲高。此是從下給臺甲高。
又以廣差乘{之|乙}高，{以|如|以[1]}本高而
一，所得，加上廣，即甲上廣。又以表差
乘乙高，如本高而一，所得，加上表，即
甲上表。其甲上廣、表即乙下廣、表。臺
上廣、表即乙上廣、表。其後求廣、表，
有增損者，皆放此。

---

1 Guo Shuchun and Liu Dun 1998, 2: 4, 21, n.
   14.

◆ *3.2.3.2. The contributions of the
two counties to the platform*

In the following, consult Figure 24, p. 65.

The method to calculate [V_B=] the volume [of earth] assigned in [cubic] *chi* and the width and length [of the part of the platform built by county B] is: Multiply the volume norm in [cubic] *chi* by the [number of] men from county B. Multiply this by the allotted number of days to obtain the volume of [the part of the platform built by county] B.

$$V_B = 75 \ chi^3/\text{man·day} \times 3{,}222 \ \text{men} \times 5 \ \text{days} = 1{,}208{,}250 \ chi^3$$

Multiply this by three and multiply by the area of [the square on] the height [h]. Divide by the difference between the upper and lower widths [$a_2 - a_1$] multiplied by the difference between the lengths [$b_2 - b_1$]. This is the *shi* [the constant term in the cubic equation].

$$shi = \frac{3h^2 V_B}{(a_2 - a_1)(b_2 - b_1)} = 146{,}802{,}375 \ chi^3$$

Multiply the height of the platform [h] by the upper width [b_1] and divide by the difference between the widths [$a_2 - a_1$] to make [K_6 =] the Height for the Upper Width.

$$K_6 = \frac{ha_1}{a_2 - a_1} = 630 \ chi \tag{3.3}$$

Multiply the height of the platform [h] by the upper length [b_1] and divide by the difference between the lengths to make [K_7 =] the Height for the Upper Length.

$$K_7 = \frac{hb_1}{b_2 - b_1} = 450 \ chi \tag{3.4}$$

Multiply the Height for the Upper Width [K_6] by the Height for the Upper Length [K_6] and multiply by 3 to make the *fangfa* [the coefficient of the linear term].

$$fangfa = 3K_6 K_7 = 850{,}500 \ chi^2$$

Add the two heights [K_6 and K_7], multiply by three, and divide by two to make the *lianfa* [the coefficient of the quadratic term].

$$lianfa = \frac{3(K_6 + K_7)}{2} = 1{,}620 \; chi$$

Extract the cube root to obtain $[h_B =]$ the height [of the part of the platform built by county] B.

$$h_B^3 + \frac{3(K_6 + K_7)}{2} h_B^2 + 3K_6 K_7 h_B = \frac{3h^2 V_B}{(a_2 - a_1)(b_2 - b_1)}$$

$$h_B^3 + 1{,}620 h_B^2 + 850{,}500 h_B = 146{,}802{,}375 \; chi^3$$

$$h_B = 135 \; chi$$

Subtract this from the full height $[h]$. The remainder is then $[h_A =]$ the height of [the part of the platform built by county] A. This is the height of [the part built by county] A from the base upwards.[1]

$$h_A = h - h_B = 45 \; chi$$

Multiply the difference between the widths $[a_2 - a_1]$ by the height [of the part of the platform built by county] B $[h_B]$ and divide by the full height $[h]$. To the result of this, add the upper width $[a_1]$. This is then $[a_3 =]$ the upper width [of the part of the platform built by county] A.

$$a_3 = \frac{(a_2 - a_1) h_B}{h} + a_1 = 85 \; chi$$

Multiply the difference between the lengths $[b_2 - b_1]$ by the height [of the part of the platform built by county] B $[h_B]$ and divide by the full height $[h]$. To the result of this, add the upper length $[b_1]$. This is then $[b_3 =]$ the upper length [of the part of the platform built by county] A.

$$b_3 = \frac{(b_2 - b_1) h_B}{h} + b_1 = 130 \; chi$$

The upper width and length of [the part built by] county A are the lower width and length of [the part built by] county B $[a_3$ and $b_3]$. The platform's upper width and length $[a_1$ and $b_1]$ are the upper width and length [of the part built by] county B.

---

1 The point of this redundant sentence is not apparent. Perhaps it is a comment by another hand.

If later [one wants to] calculate the width and length with more or fewer [counties], in all cases [the procedure] is the same.

◆ 3.2.3.3. Comment in smaller characters

In the following, consult Figure 26, p. 71.

This reflects multiplying the volume [V_B] [of the part of the platform built by county] B by six, then multiplying twice by the height of the platform [h] and dividing by the difference between the upper and lower widths [a_2-a_1] multiplied by the difference between the lengths [b_2-b_1].

$$2 \times shi = \frac{6h^2 V_B}{(a_2 - a_1)(b_2 - b_1)}$$

Further, multiplying the height of the platform [h] by the upper width [a_1] and dividing by the difference between the widths [a_2-a_1] gives [$K_6 =$] the Height for the Upper Width. Multiplying the height of the platform [h] by the upper length and dividing by the difference between the lengths [b_2-b_1] gives the Height for the Upper Length.

These are the same calculations as above, Equations (3.3) and (3.4).

Multiplying these together gives two of the small area [$PUWT$].

$$K_6 K_7 = \frac{h^2 a_1 b_1}{(a_2 - a_1)(b_2 - b_1)}$$

Further, multiplying the Height for the Lower Length [K_8] by the Height for the Upper Width [K_9] gives one of the middle areas [$PQXT$].

The comment does not define the Heights for the Lower Length and Lower Width, but they turn out to be

$$K_8 = \frac{h b_3}{b_2 - b_1}$$

$$K_9 = \frac{h a_3}{a_2 - a_1}$$

此應｛三｜六｝因乙積，臺高再乘，上下廣差乘袤差而一。又以臺高乘上廣，｛｜廣差而一，｝為上廣之高。又以臺高乘上袤，｛｜袤差而一，｝為上袤之高。｛以上廣之高乘上袤之高｜相乘｝為小冪二。｛因｜因｜又｝下袤之高｛｜｜乘上廣之高｝，為中冪一。凡下袤、下廣之高即是截高與上袤、｛與｜｜與¹｝上廣之高相連并數。然｛此｜此｜則｝有中冪定有小冪一，又有上廣之高乘截高為冪｛各｜｝一。又下廣之高乘下袤之高為大冪二。乘上袤之高為中冪一。其大冪之中｛又｜又有｜有｝小冪一，復有上廣、上袤之高｛為中冪｜｝各乘截高為｛中｜中｜｝冪各一。又截高自乘，為冪一。其中冪之內有小冪一。又上袤之高乘截高為冪一。然則截高自相乘為冪二，小冪六。又上廣上袤之高各三，以乘截高為冪六。今皆半之，故以三乘小冪。又上廣上袤之高各三，今但半之，各得一又二分之一，故三之二而一。諸冪｛｜乘｝截｛｜高｝為積尺。

1 Guo Shuchun and Liu Dun 1998, 2: 4, 21 n. 19.

Further, multiplying the Height for the Lower Length [K_8] by the Height for the Upper Width [K_9] gives one of the middle areas [$PQXT$].

$$K_6 K_8 = \frac{h^2 a_1 b_3}{(a_2 - a_1)(b_2 - b_1)}$$

The Heights for the Lower Length and the Lower Width [K_8 and K_9] are in fact the sum of the truncated height [h_B] together with the Heights for the Upper Length and Upper Width [K_6 and K_7] lined up.

$$K_8 = K_7 + h_B$$
$$K_9 = K_6 + h_B$$

Thus it has been determined that the middle area [$KLPQ$] is composed of one small area [$PUWT$] and also an area obtained by multiplying the Height for the Upper Width [K_6] by the truncated height [h_B].[1]

$$\frac{h^2 a_1 b_3}{(a_2 - a_1)(b_2 - b_1)} = K_6 K_7 + K_6 h_B$$

Further, multiplying [twice] the Height for the Lower Width [K_8] by the Height for the Lower Length [K_9] gives two of the large area [$PQRS$].

$$2 K_8 K_9 = \frac{2h^2}{(a_2 - a_1)(b_2 - b_1)} a_3 b_3$$

Multiplying . . .[2] [K_7] by the Height for the Upper Length [K_9] gives one of the middle areas [$PUVS$].

$$K_7 K_9 = \frac{h^2}{(a_2 - a_1)(b_2 - b_1)} a_3 b_1$$

Within the large area [$PQRS$] there is one small area [$PUWT$], and in addition there are the Heights for the Upper Width and the Upper Length [K_6, K_7], each multiplied by the truncated height

---

1 *Jiegao* 截高. This term is used for a different quantity in Section 3.2.3.1 above. See fn. 2, p. 130.

2 The multiplicand is missing in the text here. The mathematical context indicates that it is K_7.

$[h_B]$ to make one middle area, and the truncated height $[h_B]$ multiplied by itself to make one area.

$$2K_8K_9 = 2K_6K_7 + 2K_6h_B + 2K_7h_B + 2h_B^2$$

Within the middle area $[PUVS]$ there is one small area $[PUWT]$ together with the area obtained by multiplying the Height for the Upper Length $[K_7]$ by the truncated height $[h_B]$.

$$\frac{h^2 a_3 b_1}{(a_2 - a_1)(b_2 - b_1)} = K_7 K_9 = K_6 K_7 + K_7 h_B$$

Thus there are two of the area formed by multiplying the truncated height by itself $[h_B^2]$ and six of the small area $[K_6 K_7]$. There are also three each of the Heights for the Upper Width and Upper Length $[K_6, K_7]$.

$$2h_B^2 + 6K_6K_7 + 3K_6h_B + 3K_7h_B$$

These, multiplied by the truncated height $[h_B]$, give the six areas.[1] When all are halved, the small area [the term K_6K_7] is multiplied by three. Further, there are three each of the Heights for the Upper Width and the Upper Length $[K_6, K_7]$. When [the terms K_6h_B and K_7h_B] are halved once, in each case one and one half is obtained; this is why [these terms] are multiplied by three and divided by two. All the areas are multiplied by the truncated height $[h_B]$ to make the volumes in [cubic] *chi*.

$$h_B^3 + 3K_6K_7h_B + \tfrac{3}{2}K_6h_B^2 + \tfrac{3}{2}K_7h_B^2$$

$$= \frac{1}{2}h_B \frac{h^2}{(a_2 - a_1)(b_2 - b_1)}(2a_1b_1 + a_3b_1 + 2a_3b_3 + a_1b_3)$$

◆ *3.2.3.4. The dimensions of the ramp*

The method for calculating the width, length, and height of the ramp is: Multiply the constant volume of the uniform assessment by the two counties' 56 districts and multiply by six to make $[6W =]$ the 'Volume'.

$$W = 6{,}300 \; chi^3/\text{district} \times 56 \; \text{districts} = 352{,}800 \; chi^3$$
$$6W = 2{,}116{,}800 \; chi^3$$

求羨道廣袤高術曰：以均賦常積乘二縣五
十六鄉，又六因為積。又以道上廣多下廣
數加上廣少袤為下廣少袤。又以高多袤加
下廣少袤為下廣少高。以乘下廣少袤為
隅陽冪。又以下廣少上廣乘之，為鱉隅{｜
積}。以減積，餘，三而一，為實。幷下
廣少袤與下廣少高，以下廣少上廣乘之，
為鱉從橫廉冪。三而一，加隅{‖陽[1]}冪，
為方法。又以三除上廣多下廣，以下廣少
袤、下廣少高加之，為廉法，從。開立方
除之，即下廣。加廣差即上廣，加袤多上
廣於上廣即袤，加{廣|高}多袤即道高。

1 Guo Shuchun and Liu Dun 1998, 2: 4, 20 n. 26.

1 Here we correct an error in Lim and Wagner 2013a: 15.

In the following, consult Figure 24, p. 65.

Add the difference between the upper width and the lower width of the ramp $[c_1-c_2]$ to the difference between the upper width and the length $[l-c_1]$ to obtain the difference between the length and the lower width $[l-c_2]$.

$$l-c_2 = (c_1-c_2) + (l-c_1) = 116 \ chi$$

Add the difference between the height and the length $[g-l]$ to the difference between the length and the lower width $[l-c_2]$ to make the difference between the height and the lower width $[g-c_2]$.

$$g-c_2 = (g-l) + (l-c_2) = 156 \ chi$$

Multiply this by the difference between the length and the lower width $[l-c_2]$ to make $[K_{11} =]$ the Area for the Corner $Yang[ma]$.

$$K_{11} = (g-c_2)(l-c_2) = 18{,}096 \ chi^2$$

Multiply by the difference between the upper width and the lower width $[c_1-c_2]$ to make $[K_{12} =]$ the Volume for the $Bie[nao]$ Corner.

$$K_{12} = K_{11}(c_1-c_2) = 217{,}152 \ chi^3$$

Subtract this from the 'Volume' $[6W]$ and divide the difference by three to make the shi [the constant term of the cubic equation].

$$shi = \frac{6W - K_{12}}{3} = 633{,}216 \ chi^3$$

Add the difference between the length and the lower width $[l-c_2]$ and the difference between the height and the lower width $[g-c_2]$ and multiply by the difference between the upper width and the lower width $[c_1-c_2]$ to make $[K_{13} =]$ the Area for the Vertical and Horizontal Edge $Bie[nao]$.

$$K_{13} = ([l-c_2] + [g-c_2])(c_1-c_2) = 3{,}264 \ chi^2$$

Divide by three and add the Area for the Corner $[Yangma]$ $[K_{11}]$ to make the $fangfa$ [the coefficient of the linear term].

$$fangfa = \frac{K_{13}}{3} + K_{11} = 19{,}184 \ chi^2$$

Divide the difference between the upper width and the lower width $[c_1 - c_2]$ by three and add to this the difference between the length and the lower width $[l - c_2]$ and the difference between the height and the lower width $[g - c_2]$ to make the *lianfa* [the coefficient of the quadratic term].

$$lianfa = \frac{c_1 - c_2}{3} + (l - c_2) + (g - c_2) = 276 \; chi$$

Extract the cube root to obtain the lower width $[c_2]$.

$$c_2^3 + \left[\frac{c_1 - c_2}{3} + (l - c_2) + g - c_2 \right] c_2^2 + \left(\frac{K_{13}}{3} + K_{11} \right) c_2$$

$$= \frac{6W - K_{12}}{3}$$

$$c_2^3 + 276\, c_2^2 + 19{,}184\, c_2 = 633{,}216 \; chi^3$$

$$c_2 = 24 \; chi.$$

Add this $[c_2]$ to the difference in the widths $[c_1 - c_2]$ to make the upper width $[c_1]$. Add the difference between the length and the upper width $[l - c_1]$ to the upper width $[c_1]$ to make the length $[l]$. Add the difference between the height and the length $[g - l]$ to make the the height $[g]$ of the ramp.

$$c_1 = c_2 + (c_1 - c_2) = 36 \; chi$$

$$l = c_1 + (l - c_1) = 140 \; chi$$

$$g = l + (g - l) = 180 \; chi \hspace{2cm} (3.5)$$

◆ *3.2.3.5. The contributions of the*
two counties to the ramp

求羨道均給積尺，甲縣受廣、表，術曰：
以均賦常積乘甲縣一十三鄉，又六因為
積。以表再乘之，以道上下廣差乘臺高為
法而一，為實。又三因下廣，以表乘之，
如上下廣差而一，為都廉，從。開立方除
之，即甲表。以廣差乘甲表，本表而一，
以下廣加之，即甲上廣。又以臺高乘甲
表，本表除之，即甲高。

The method to calculate the volume [of earth] assigned in [cubic] *chi* and the width and length of [the part of the ramp supplied by] county A is: Multiply the constant volume of the uniform assessment by county A's 13 districts and multiply this by six to make $[6W_A =]$ the 'Volume'.

$$W_A = 6{,}300 \; chi^3/\text{district} \times 13 \; \text{districts} = 81{,}900 \; chi^3$$

$$6W_A = 491{,}400 \; chi^3$$

Multiply this twice by the length $[l]$. Multiply the difference between the upper and lower widths $[c_1 - c_2]$ of the ramp by the height of the platform $[h = g]$[1] to make the divisor, and divide to make the *shi* [the constant term of the cubic equation].

$$shi = \frac{6W_A l^2}{(c_1 - c_2)g} = 4,459,000 \ chi^3$$

Multiply the lower width $[c_2]$ by three, multiply by the length $[l]$, and divide by the difference between the upper and lower width $[c_1]$ to make the *dulian* 都廉 [the coefficient of the quadratic term].[2]

$$dulian = \frac{3c_2 l}{c_1 - c_2} = 840 \ chi$$

Extract the cube root to obtain $[l_A =]$ the length [of the part of the ramp built by county] A.

$$l_A^3 + \frac{3c_2 l}{c_1 - c_2} l_A^2 = \frac{6W_A l^2}{(c_1 - c_2)g}$$

$$l_A^3 + 840 \ l_A^2 = 4,459,000 \ chi^3$$
$$l_A = 70 \ chi$$

Multiply the difference in widths $[c_1 - c_2]$ by the length $[l_A]$ of [the part built by county] A. Divide by the full length $[l]$ and add the lower width $[c_2]$ to make $[c_3 =]$ the upper width [of the part built by county] A.

$$c_3 = c_2 + \frac{c_1 - c_2}{l} l_A = 30 \ chi$$

Multiply the height of the platform $[h = g]$ by the length $[l_A]$ [of the part built by county] A and divide by the full length $[l]$. This is then $[g_A =]$ the height [of the part built by county] A.

$$g_A = \frac{g}{l} l_A = 90 \ chi$$

---

1   The height of the ramp, *g*, was calculated in Section 3.2.3.4 above (Equation (3.5)), and it turned out to be the same as the height of the platform, $g = h = 180 \ chi$. It is therefore not important, but something of a surprise, that the text refers here to the height of the *platform* rather than that of the *ramp*.

2   Wang Xiaotong's usual terms for the quadratic and linear coefficients are respectively *lianfa* 廉法 and *fangfa* 方法. Here and in a few other places in the text, however, he uses the terms *dulian* 都廉 and *yuanfang* 垣方 for these (Sections 3.3.3.3, 3.4.4, 3.4.4.3, and 3.5.3.2). We have been unable to determine whether the terms are exact synonyms, or differ in usage in some way.

3.3. Problem 3:
Construction of a dyke

3.3.1. The problem

In the following, consult Section 2.6.1 above and Figures 30 and 33, pp. 82 and 87. From the calculations in Section 3.3.3.2 below (equations (3.7) and (3.8), pp. 144 and 145), two implicit assumptions become apparent: that the planes *ABCD* and *EFGH* are perpendicular to *CDEF*, and that $a_1 = a_2$. The text sometimes refers to the 'eastern' and 'western' upper widths (a_1 and a_2), sometimes simply to the 'upper width', which here will be denoted *a* (defined to be equal to a_1 and a_2).

假令築隄，西頭上、下廣差六丈八尺二寸，東頭上、下廣差六尺二寸，東頭高少於西頭高三丈一尺，上廣多東頭高四尺九寸，正袤多於東頭高四百七十六尺九寸。甲縣六千七百二十四人，乙縣一萬六千六百七十七人，丙縣一萬九千四百四十八人，丁縣一萬二千七百八十一人。四縣每人一日穿土九石九斗二升。每人一日築常積一十一尺四寸、十三分寸之六。穿方一尺得土八斗。古人負土二斗四升八合，平道行一百九十二步，一日六十二到。今隔山渡水取土，其平道只有一十一步，山斜高三十步，水寬一十二步，上山三當四，下山六當五，水行一當二，平道踟蹰十加一，載輸一十四步。減計一人作功為均積，四縣共造，一日役畢。今從東頭與甲，其次與乙、丙、丁。問：給斜、正袤，與高，及下廣，并每人一日自穿、運、築程功，及隄上、下高、廣各幾何？

Suppose that a dyke is to be built. The difference between the upper and lower widths of the western end is [$b_2 - a_2 =$] 6 *zhang* 8 *chi* 2 *cun* [= 682 *cun*], the difference between the upper and lower widths of the eastern end is [$b_1 - a_1 =$] 6 *chi* 2 *cun* [= 62 *cun*], the height of the eastern end is [$h_2 - h_1 =$] 3 *zhang* 1 *chi* [= 310 *cun*] less than the height of the western end, the upper width is [$a - h_1 =$] 4 *chi* 9 *cun* [= 49 *cun*] greater than the height of the eastern end, the straight length is [$l - h_1 =$] 476 *chi* 9 *cun* [= 4,769 *cun*] greater than the height of the eastern end.

County A [sends] 6,724 workers, county B [sends] 16,677 workers, county C [sends] 19,448 workers, and county D [sends] 12,781 workers. Each person from the four counties can in one day excavate 9 *dan* 9 *dou* 2 *sheng* [= 9,920 *ge*] of soil. Each person can build a constant volume of 11 *chi* 4 $^6/_{13}$ *cun* [i.e. 114 $^6/_{13}$ *cun*^{3}] per day. Digging out 1 [cubic] *chi* of soil results in 8 *dou* [= 800 *ge*] of soil.

People in former times, carrying 2 *dou* 4 *sheng* 8 *ge* [= 248 *ge*] of soil on their backs and travelling 192 *bu* on a level road, did 62 trips in one day. In the present situation there are hills to climb and rivers to cross to obtain the soil: there are only 11 *bu* of level road, the slanted height of the hill is 30 *bu*, and the width of the river is 12 *bu*. When climbing a hill 3 [*bu*] is equivalent to 4 [*bu* of level road], when descending a hill 6 [*bu*] is equivalent to 5 [*bu*], and when crossing water 1 [*bu*] is equivalent to 2 [*bu*].

For rest on a level road, one is added for every ten [*bu*]. Loading and unloading [is equivalent to transportation of] 14 *bu*.

In the calculations, one man's work is simplified to an equal volume.

> That is, it is assumed for the purpose of the calculation that each man does an equal share of each of the phases of the work: digging, transporting, and building.

The four counties build the dyke together, and the work is completed in one day. Starting from the eastern end, [the work is] assigned to county A, then consecutively to B, C, and D.

What is the contribution [of each county] with regard to the slanted and straight lengths and the height and lower width [of the parts of the dyke built by the counties], and how much soil does each person dig, transport and build? What are the upper and lower heights and widths of the dyke?

3.3.2. Answer

The final amount that one man digs, transports and builds in one day is 4 *chi* 9 *cun* 6 *fen* [i.e. 4.96 *chi*^{3} = 4,960 *cun*^{3}].

Western end:

> height: [h_2 =] 3 *zhang* 4 *chi* 1 *cun* [= 341 *cun*];
> upper width: [a_2 =] 8 *chi*; [= 80 *cun*];
> lower width: [b_2 =] 7 *zhang* 6 *chi* 2 *cun* [= 762 *cun*].

Eastern end:

> height: [h_1 =] 3 *chi* 1 *cun* [= 31 *cun*];
> upper width: [a_1 =] 8 *chi* [= 80 *cun*];
> lower width: [b_1 =] 1 *zhang* 4 *chi* 2 *cun* [= 142 *cun*];
> straight length: [l =] 48 *zhang* [= 4,800 *cun*];
> slanted length: [s =] 48 *zhang* 1 *chi* [= 4,810 *cun*].

County A:

> straight length: [l_A =] 19 *zhang* 2 *chi* [= 1,920 *cun*];
> slanted length: [s_A =] 19 *zhang* 2 *chi* 4 *cun* [= 1,924 *cun*];
> lower width: [b_{2A} =] 3 *zhang* 9 *chi* [= 390 *cun*];
> height: [h_A =] 1 *zhang* 5 *chi* 5 *cun* [= 155 *cun*].

County B:

> straight length: [l_B =] 14 *zhang* 4 *chi* [= 1,440 *cun*];
> slanted length: [s_B =] 14 *zhang* 4 *chi* 3 *cun* [= 1,443 *cun*];
> lower width: [b_{2B} =] 5 *zhang* 7 *chi* 6 *cun* [= 576 *cun*];
> height: [h_B =] 2 *zhang* 4 *chi* 8 *cun* [= 248 *cun*].

答曰：

一人一日自穿、運、築，程功四尺九寸
{二|六}分。

西頭高三丈四尺一寸，
　上廣八尺，
　下廣七丈六尺二寸。

東頭高三尺一寸，
　上廣八尺，
　下廣一丈四尺二寸，
　正袤四十八丈，
　斜袤四十八丈一尺。

甲縣正袤一十九丈二尺，
　斜袤一十九丈二尺四寸，
　下廣三丈九尺，
　高一丈五尺五寸。

乙縣正袤一十四丈四尺，
　斜袤一十四丈四尺三寸，
　下廣五丈七尺六寸，
　高二丈四尺八寸。

丙縣正袤九丈六尺，
　斜袤九丈六尺二寸，
　下廣七丈，
　高三丈一尺。

丁縣正袤四丈八尺，
　斜袤四丈八尺一寸，
　下廣七丈六尺二寸，
　高三丈四尺一寸。

County C:

> straight length: [l_C =] 9 *zhang* 6 *chi* [= 960 *cun*];
> slanted length: [s_C =] 9 *zhang* 6 *chi* 2 *cun* [= 962 *cun*];
> lower width: [b_{2C} =] 7 *zhang* [= 700 *cun*];
> height: [h_C =] 3 *zhang* 1 *chi* [= 310 *cun*].

County D:

> straight length: [l_D =] 4 *zhang* 8 *chi* [= 480 *cun*];
> slanted length: [s_D =] 4 *zhang* 8 *chi* 1 *cun* [= 481 *cun*];
> lower width: [$b_{2D} = b_2$ =] 7 *zhang* 6 *chi* 2 *cun* [= 762 *cun*];
> height: [$h_D = h_2$ =] 3 *zhang* 4 *chi* 1 *cun* [= 341 *cun*].

3.3.3. Method

◆ 3.3.3.1. The volume of the dyke

The method for calculating the number of trips [of one man], the final amount of work and how many [cubic] *chi* [each man] digs, moves and builds is:

Lay out [the following on the counting board:] to ascend the mountain: 40 *bu*, to descend the mountain: 25 *bu*, to cross the river: 24 *bu*, to travel on the even road: 11 *bu*, for the time to rest: for each 10 add 1, for loading and unloading: 14 *bu*. For each man one round-trip amounts to 124 *bu*.

The length of the road from excavation to building site and back is equivalent to

$$\left(11 + 30 \times \frac{4}{3} + 30 \times \frac{5}{6} + 12 \times \frac{2}{1}\right) \times 1.1 + 14 = 124 \; bu$$

Given that in ancient times men carried 2 *dou* 4 *sheng* 8 *ge* [= 248 *ge*] on their backs and traveled 192 *bu* on an even road, multiply [this] by 62 trips to obtain the dividend. Then take the number of *bu* for one round trip as the divisor. Divide to obtain the number of trips for each man moving soil.

$$\frac{192 \; bu \,/\, \text{man} \cdot \text{day} \; \times 62 \; \text{trips} \,/\, \text{man} \; \times 1 \; \text{day}}{124 \; bu \,/\, \text{man}}$$

$$= 96 \; \text{round trips per man}$$

Multiply by the amount of soil carried on the back in one trip [248 *ge*]. Then divide by the [measure of] dug out [loose soil equivalent to] 1 [cubic] *chi* [800 *ge/chi*^{3}], obtaining the volume of soil that one man transports per day.

求人到程功，運、築積尺術曰：置上山四十步，下山二十五步，渡水二十四步，平道二十一步，跔躇之間十加一，載輸一十四步，一返計一百二十四步。以古人負土二斗四升八合，平道行一百九十二步，以乘一日六十二到為實。卻以一返步為法除，得自運土到數也。又以一到負土數乘之，卻以穿方一尺土數除之，得一人一日運功積。又以一人穿土九石九斗二升，以穿方一尺土數除之為法，除之，得穿用人數。復置運功積，以每人一日常積除之，得築用人數。并之得六人，共成二十九尺七寸六分。以六人除之，即一人程功也。

$$\frac{248 \ ge \ / \ \mathrm{man} \cdot \mathrm{day}}{800 \ ge \ / \ chi^3} \times 96 = 29.76 \ chi^3 \ / \ \mathrm{man} \cdot \mathrm{day}$$

Divide the 9 *dan* 9 *dou* 2 *sheng* [= 9,920 *ge*] of soil that one man digs per day by the [measure of] dug out [loose soil equivalent to] 1 [cubic] *chi* [800 *ge*/*chi*³] to make the divisor; divide [the amount of soil one man can transport per day] by this to obtain the number of men employed for digging soil.

$$\frac{800 \ ge \ / \ chi^3 \times 29.76 \ chi^3}{9920 \ ge \ / \ \mathrm{man}} = 2.4 \ \mathrm{men}$$

Again lay out the volume of soil transported per day [29.76 *chi*³] and divide by the constant volume of soil that each person [can tamp] per day [11 $^{29}/_{65}$ *chi*³] to obtain the number of men employed to tamp the soil.

$$\frac{29.76 \ chi^3}{11^{29}/_{65} \ chi^3 \ / \ \mathrm{man}} = 2.6 \ \mathrm{men}$$

Add these to obtain [1 + 2.4 + 2.6 =] 6 men. Together they produce 29 *chi* 7 *cun* 6 *fen* [i.e. 29.76 *chi*³] [of tamped soil per day]. Divide this by the 6 men; this is one man's amount of work.

The contribution to the volume of the dyke by one man-day of labour is

$$\frac{29.76 \ chi^3}{6 \ \mathrm{men}} = 4.96 \ chi^3 \ / \ \mathrm{man} = 4,960 \ cun^3 \ / \ \mathrm{man} \quad (3.6)$$

◆ *3.3.3.2. The dimensions of the dyke*

The method for calculating the upper and lower widths and the height and length of the dyke is: Multiply one man's amount of work in one day by the total number of men to make the volume of the dyke.

Using (3.6), the volume of the dyke is

$$V = 4,960 \ cun^3 \ / \ \mathrm{man} \times 55,630 \ \mathrm{men} = 275,924,800 \ cun^3.$$

Multiply the difference between the heights [h_2-h_1] by the difference between the lower widths [b_2-b_1] and divide by 6 to obtain [$K_1 =$] the Area for the *Bienao*.

求隄上、下廣及高、袤術曰：
一人一日程功乘總人為隄積。以高差乘下廣差，六而一，為鼈幂。又以高差{|乘}小頭廣差，二而一，為大臥塹頭幂。又半高差乘上廣多東頭高之數，為小臥塹頭幂。并三幂，為大小塹鼈率。乘正袤多小高之數，以減隄積，餘為實。又置半高差，及半小頭廣差與上廣多小頭高之數，并三差，以乘正袤多小頭高之數。以加率為方法。又并正袤多小高、{并|}上廣多小高及半高差，{而增之|}兼半小頭廣差加之為廉法，從。開立方除之，即小高。加差即各得廣、袤、高。又正袤自乘、高差自乘，并，而開方除之，即斜袤。

$$b_2-b_1 = (b_2-a) - (b_1-a) = 620 \; cun$$

$$K_1 = \frac{(h_2-h_1)(b_2-b_1)}{6}$$

$$= \frac{(h_2-h_1)\left[(b_2-a)-(b_1-a)\right]}{6} = 32{,}033\tfrac{1}{3} \; cun^3 \qquad (3.7)$$

Multiply the difference between the heights [h_2–h_1] by the difference between the widths of the small end [b_1–a_1] and divide by 2 to make [K_2 =] the Area for the End of the Large Recumbent *Qiandu*.

$$K_2 = \frac{(h_2-h_1)(b_1-a_1)}{2} = 9{,}610 \; cun^3$$

Further, halve the product of the difference between the heights [h_2–h_1] and the difference between the upper width and the height at the eastern end [a–h_1] to make [K_3 =] the Area for the End of the Small Recumbent *Qiandu*.

$$K_3 = \frac{(h_2-h_1)(a-h_1)}{2} = 7{,}595 \; cun^2$$

Add the three areas to make [K_4 =] the Proportion for the Large and Small *Qiandu* and *Bienao*.

$$K_4 = K_1+K_2+K_3 = 49{,}238\tfrac{1}{3} \; cun^2$$

Multiply by the difference between the straight length and the small height [l–h_1] and subtract [the result] from the volume of the dyke [V]. The difference is the *shi* [the constant term of the cubic equation].

$$shi = V-K_4(l-h_1) = 41{,}107{,}188\tfrac{1}{3} \; cun^3$$

Further, lay out half of the difference between the heights [h_2–h_1], half of the difference between the widths of the small end [b_1–a_1], and the difference between the upper width and the height of the small end [a–h_1]. Add together the three differences and multiply by the difference between the straight length and the height of the small end [l–h_1]. Add this to the Proportion [K_4] to obtain the *fangfa* [the coefficient of the linear term].

$$fangfa = \left[\frac{h_2-h_1}{2} + \frac{b_1-a_1}{2} + (a-h_1)\right](l-h_1) + K_4$$

$$= 1,169,953\tfrac{1}{3}\ cun^2$$

Add together the difference between the straight length and the small height $[l-h_1]$, the difference between the upper width and the small height $[a-h_1]$, and half of the difference between the heights $[h_2-h_1]$. At the same time add half of the difference between the widths of the small end $[b_1-a_1]$ to this to obtain the *lianfa* [the coefficient of the quadratic term].

$$lianfa = \left(l-h_1\right)+\left(a-h_1\right)+\frac{h_2-h_1}{2}+\frac{b_1-a_1}{2}=5,004\ cun$$

Extract the cube root; the result is the small height $[h_1]$.

$$h_1^3 + \left[\left(l-h_1\right)+\left(a-h_1\right)+\frac{h_2-h_1}{2}+\frac{b_1-a_1}{2}\right]h_1^2$$
$$+\left\{\left[\frac{h_2-h_1}{2}+\frac{b_1-a_1}{2}+\left(a-h_1\right)\right]\left(l-h_1\right)+K_4\right\}h_1 \qquad (3.8)$$
$$=V-K_4\left(l-h_1\right)$$

$$h_1^3 + 5,004\ h_1^2 + 1,169,953\tfrac{1}{3}\ h_1 = 41,107,188\tfrac{1}{3}\ cun^3$$
$$h_1 = 31\ cun$$

Add this to the differences to obtain the widths, length and height.

$$a = a_1 = a_2 = (a-h_1) + h_1 = 80\ cun$$
$$b_1 = (b_1-a_1) + a_1 = 142\ cun$$
$$b_2 = (b_2-a_2) + a_2 = 762\ cun$$
$$h_2 = (h_2-h_1) + h_1 = 341\ cun$$
$$l = (l-h_1) + h_1 = 4,800\ cun$$

Add together the product of the straight length by itself $[l]$ and the product of the difference in heights by itself $[h_2-h_1]$ and extract the square root. This is the slanted length $[s]$.

$$s = \sqrt{l^2 + \left(h_2-h_1\right)^2} = 4,810\ cun$$

◆ *3.3.3.3. The contributions of the counties*

In the following, consult Figure 33, p. 87.

求甲縣高、廣、正、斜表術曰：以程功乘
甲縣人，以六因取積，又乘表冪。以下廣
差乘高差｛以｜為｜以｝法除之，為實。又并
小頭上、下廣，以乘小高，三因之為垣頭
冪。又乘表冪，如法而一，為垣方。又三
因小頭下廣，以乘正表，以廣差除之，為
都廉，從。開立方除之，得小頭｛｜表｜｝，
即甲表。又以下廣差乘之，｛所得｜｜所得｝
以正表除之，所得，加東頭下廣即甲廣。
又以兩頭高差乘甲表，以正表除之，以加
東頭高，即甲高。又以甲表自乘，以隁東
頭高減甲高，餘，自乘，并二位，以開方
除之，即得斜表。｛求高廣以本表及高廣
差求之｜｝若求乙、丙、丁，各以本縣人功
積尺，每以前大高、廣為後小高、廣。凡
廉母自｛來｜乘｝為方母，廉母乘方母為實
母。

The method for calculating the height, the width, and the straight and slanted lengths of [the part of the dyke built by] county A is: Multiply the regulation amount of work by the number of men in county A.

$$V_A = 4{,}960 \; cun^3 \, / \, man \times 6{,}724 \; men = 33{,}351{,}040 \; cun^3$$

Multiply the volume obtained [V_A] by 6. Multiply this by the area of [the square on] the length [l]. Multiply the difference between the lower widths [b_2-b_1] by the difference between the heights [h_2-h_1] to make [$K_5 =$] the Divisor and divide to make the *shi* [the constant term in the cubic equation].

$$K_5 = (b_2-b_1)(h_2-h_1) = 192{,}200 \; cun^2$$

$$shi = \frac{6V_A l^2}{K_5} = 23{,}987{,}761{,}548^{12}/_{31} \; cun^3$$

Further add together the upper and lower widths of the small end [a_1, b_1] and multiply by the small height [h_1]. Multiply this by three to make [$K_6 =$] the Area for the End Wall.

$$K_6 = 3\left(a_1+b_1\right)h_1 = 20{,}646 \; cun^2$$

Multiply by the area of [the square on] the length [l] and divide by the Divisor [K_5] to obtain the *yuanfang* [the coefficient of the linear term].

$$yuanfang = \frac{K_6 l^2}{K_5} = 2{,}474{,}941^{29}/_{31} \; cun^2$$

Triple the lower width of the small end [b_1] and multiply by the straight length [l]. Divide by the difference in the widths [b_2-b_1] to make the *dulian* [the coefficient of the quadratic term].

$$dulian = \frac{3b_1 l}{b_2-b_1} = 3{,}298^2/_{31} \; cun$$

Extract the cube root to obtain the length at the small end [l_A], which is the length of [the contribution of] county A.

$$l_A{}^3 + \frac{3b_1 l}{b_2-b_1}l_A{}^2 + \frac{K_6 l^2}{K_5}l_A = \frac{6V_A l^2}{K_5}$$

$$l_A^3 + 3{,}298\,^2/_{31}\, l_A^2 + 2{,}474{,}941\,^{29}/_{31}\, l_A$$

$$= 23{,}987{,}761{,}548\,^{12}/_{31}\ cun^3$$

$$l_A = 1{,}920\ cun$$

Further multiply this by the difference between the lower widths $[b_2-b_1]$ and divide by the straight length $[l]$. To the result add the lower width of the eastern end $[b_1]$. This is the width $[b_{2A}]$ of [the contribution of] county A.

Multiply the difference between the heights of the two ends $[h_2-h_1]$ by the length $[l_A]$ of [the contribution of] county A. Divide by the straight length $[l]$ and add the height of the eastern end $[h_1]$. This is the height $[h_A]$ of [the contribution of] county A.

$$b_{2A} = \frac{l_A\left(b_2 - b_1\right)}{l} + b_1 = 390\ cun$$

$$h_A = \frac{\left(h_2 - h_1\right)l_A}{l} + h_1 = 155\ cun$$

Further multiply the length $[l_A]$ of [the contribution of] county A by itself. Subtract the height $[h_1]$ of the eastern end of the dyke from the height $[h_A]$ of [the contribution of] county A. Multiply the difference by itself. Add the two quantities and extract the square root to obtain the slanted length $[s_A]$.

$$s_A = \sqrt{l_A^2 + \left(h_A - h_1\right)^2} = 1{,}924\ cun$$

In calculating [the contributions of] counties B, C, and D, use the number of men in the respective county and the regulation amount of manual work in [cubic] *chi* [that one man can do] [to obtain the volume]. For each [county] let the previous large height and width be the later small height and width.

In each case the denominator of the *fang[fa]* [the linear coefficient], multiplied by itself, is the denominator of the *lian[fa]* [the quadratic coefficient], and the denominator of the *fang[fa]* multiplied by the denominator of the *lian[fa]* is the denominator of the *shi* [the constant term].

This sentence describes the normalization of the fractions involved in the cubic equation. See Section 1.5 above.

◆ *3.3.3.4. Comment in smaller characters*

The text of the comment is clearly corrupt. Elsewhere in this book we have almost always silently accepted the emendations of Qian Baocong, but here we follow the original text and discuss each of Qian Baocong's emendations. In Section 2.6.6 above we suggest one possible reconstruction of the original intention of the comment.

In the following, consult Figure 34, pp. 90–91.

In this case there is a level dyke [*DCKJHEFG*] at the top and a *yanchu* [*ABKJHG*] at the bottom. The difference between the two heights is then the height of the *[yan]chu*. The *[yan]chu* has on each side a *bienao* [*AJLH* and *BMKG*] and in the middle a *qiandu* [*JKMLHG*].[1]

Define

$$V_{\text{level dyke}} = \text{volume of } DCKJHEFG = \frac{(a_1 + b_1)h_1 l}{2} = \frac{K_6 l}{6}$$

$$V_{qiandu} = \text{volume of } JKMLHG = \frac{b_1(h_2 - h_1)l}{2}$$

$$V_{bienao} = \text{sum of the volumes of } AJLH \text{ and } BMKG$$

$$= \frac{(b_2 - b_1)(h_2 - h_1)l}{6}$$

$$V_{\text{A level dyke}} = \text{volume of } NPRQHEFG$$

$$V_{\text{A } qiandu} = \text{volume of } QRUTHG$$

$$V_{\text{A } bienao} = \text{sum of the volumes of } STQH \text{ and } UVRG$$

then the comment that follows appears to use the following equalities:

$$\frac{V_{\text{A level dyke}}}{V_{\text{level dyke}}} = \frac{l_{\text{A}}}{l}$$

此平隄在上，羨除在下。兩高之差即除高。其{餘|除}兩邊各一鱉腝，中一塹堵。今以袤再乘{積|積|六因積[1]}，廣差乘{袤|高}差而一，得截鱉腝袤，再{|自|}乘為立方一。又塹堵袤自乘為羃{三|一}，又三因小頭下廣，大袤乘之，廣差而一，與羃為高，故為廉法。又并小頭上下廣又三之，{|以乘小頭高為頭羃，}意同六除。然此頭羃，本乘截袤。又袤{|再}乘之，差相乘而一。今還依數乘除{一|}頭羃為從。{|開立方除之}，得截袤{為廣|}。

1 Li Huang 1832, *shang* 上, 48a.

1 Note that the comment always gives both characters for *bienao* 鱉腝 and *qiandu* 塹堵, while the main text consistently abbreviates these with their first characters, *bie* and *qian*. And the comment abbreviates *yanchu* 羨除 with its second character, *chu*, in all but the first occurrence of the word.

$$\frac{V_{A\,bienao}}{V_{bienao}} = \left(\frac{l_A}{l}\right)^3$$

$$\frac{V_{A\,qiandu}}{V_{qiandu}} = \left(\frac{l_A}{l}\right)^2$$

When [6 times][1] the volume [of the two *bienao*] is multiplied twice by the length [*l*] and divided by the product of the difference between the widths [b_2-b_1] and the difference between the heights [h_2-h_1], is obtained the length of the truncated *bienao* [l_A] multiplied twice [by itself] to make one cube.

> With the indicated additions by Li Huang and Qian Bao-cong ('six times', 'of the two *bienao*', 'by itself'), the text states that

$$l_A{}^3 = \frac{6l^2 V_{A\,bienao}}{(b_2-b_1)(h_2-h_1)} \qquad (3.9)$$

Further, the product of the length of the *qiandu* by itself makes (three / one) area(s).

> Qian Baocong emends 'three' to 'one'.
>
> The point of the sentence is obscure. Is the 'length' that of the whole *qiandu*, *l*, or that of the truncated *qiandu*, l_A?

Further, multiplying the lower width of the smaller face [b_1] by three, multiplying by the longer length [*l*], and dividing by the difference between the widths [b_2-b_1] makes the height corresponding to the area(s).

> This calculation is

$$\frac{3b_1 l}{b_2 - b_1}$$

> and there is no obvious geometric interpretation for the statement that this quantity 'corresponds to' the area

---

1   Li Huang (1832, *shang* 上, 48a) inserts *liu yin* 六因; Qian Baocong (1966: 44, fn. 2) suggests that this emendation is not necessary, since the volume V_A is multiplied by 6 in the main text, at the beginning of Section 3.3.3.3 above. However, he understands the underlying mathematics in the same way as we do, that is, that the 'volume' is that of the two *bienao*, and that the sentence expresses (3.9).

mentioned immediately before it, l^2 or $l_A{}^2$. If the area is taken to be $l_A{}^2$, the following sentence makes some sense.

This is therefore the *lianfa* [the coefficient of the quadratic term].

$$lianfa = \frac{3b_1 l}{b_2 - b_1}$$

Further, adding together the upper and lower widths of the smaller face [a_1 and b_1] and multiplying by three, [. . .]

> Something is clearly missing here. Qian Baocong, following Li Huang, inserts, 'and multiplying by the height of the small face [h_1] gives the area of the face,' so that the calculation would be $3(a_1+b_1)h$, which gives 6 times the area of the 'face', *EFGH*.

is equivalent to (dividing by six / six of the *yanchu*).

> *Chu* 除 in this sentence might mean 'divide', or it might be the abbreviation for *yanchu*. The meaning of the sentence is obscure under either interpretation.

This face area is to be multiplied by the truncated length [l_A].

> This multiplication would give $6V_\text{Aleveldyke}$.

Further multiplying [twice][1] by the length [presumably l rather than l_A] and dividing by the product of the differences [$(b_2-b_1)(h_2-h_1)$] [. . .]

> Something appears again to be missing. With Qian Baocong's addition of 'twice', the calculation is

$$\frac{3(a_1+b_1)h_1 l_A l^2}{(b_2-b_1)(h_2-h_1)} = \frac{K_6 l^2}{K_5} l_A$$

Now returning to the *yishu* 依數 [perhaps the coefficients of the cubic equation?] and multiplying [. . .] and dividing [. . .] one area of the face [. . .] makes [. . .] *zong/cong* 從 [extract the cube root] [. . .] obtain the truncated length [l_A] [. . .] is the width.

> This sentence is surely corrupt, but seems to be concerned with the normalization of the cubic equation. Normalization would involve multiplying by l^2 and di-

---

1 Qian Baocong (1966a) adds *zai* 再.

viding by $(b_2-b_1)(h_2-h_1)/6$; the latter is the sum of the areas of *ALJ* and *BMK*, the 'faces' of the two *bienao*.

◆ *3.3.3.5. The volume of a dyke*

Method of calculating the volume of a dyke:

Lay out the height of the western end [h_2], double it, and add the height of the eastern end [h_1]. Further add together the upper and lower widths of the western end [a_2, b_2]. Halve this and multiply [the previous result] by it.

Further lay out the height of the eastern end [h_1], double it, and add the height of the western end [h_2]. Further add together the upper and lower widths of the eastern end [a_1, b_1]. Halve this and multiply [the previous result] by it.

Add together the two quantities and multiply by the straight length [l]. Divide by 6 to obtain the volume of the dyke.

$$V = \frac{l}{6}\left((2h_2+h_1)\frac{a_2+b_2}{2}+(2h_1+h_2)\frac{a_1+b_1}{2}\right) \qquad (3.10)$$

求隄都積術曰：

置西頭高倍之，加東頭高，又幷西頭上、下廣，半而乘之。又置東頭高倍之，加西頭高，又幷東頭上、下廣，半而乘之。幷二位積，以正袤乘之，六而一，得隄積也。

3.4. Problem 4: Construction of a 'dragon tail' dyke

3.4.1. The problem

In the following, consult Section 2.7 above and Figure 39, p. 97.

假令築龍尾隄，其隄從頭高、上闊以次低狹至尾。上廣多，下廣少。隄頭上下廣差六尺，下廣少高一丈二尺，少表四丈八尺。甲縣二千三百七十五人，乙縣二千三百七十八人，丙縣五千二百四十七人。各人程功常積一尺九寸八分。一日役畢。三縣共築。今從隄尾與甲縣，以次與乙、丙。問龍尾隄從頭至尾高、表、廣，及各縣別給高、表、廣各多少？

Suppose that a dragon-tail dyke is to be built. The dyke is high at the face and the top is broad, [while the dyke] successively becomes lower and narrower until the tail is reached. The upper width [c_1] is larger and the lower width [c_2] is smaller. The difference between the upper and lower widths of the face of the dyke is [$c_1 - c_2 =$] 6 chi, the lower width is [$g - c_2 =$] 1 zhang 2 chi [$= 12$ chi] shorter than the height, and [$l - c_2 =$] 4 zhang 8 chi [$= 48$ chi] shorter than the length. County A [sends] 2,375 men, county B [sends] 2,378 men, and county C [sends] 5,247 men. Each man's contribution is a constant volume of 1 chi 9 cun 8 fen [i.e. 1.98 chi³]. In one day the work is finished.

The three counties build [the dyke] together. From the end of the dyke, [the work is] given to county A and in turn given to county B and C. From the end to the tail of the dragon tail dyke, how high, long and wide is it, and what is the contribution in height, length and width of each county?

3.4.2. Answer

答曰：
　高三丈，
　　上廣二丈四尺，
　　下廣一丈八尺，
　　表六丈六尺。
　甲縣高一丈五尺，
　　表三丈三尺，
　　上廣二丈一尺。
　乙縣高二丈一尺，
　　表一丈三尺二寸，
　　上廣二丈二尺二寸。
　丙縣高三丈，
　　表一丈九尺八寸，
　　上廣二丈四尺。

Height: [$g =$] 3 zhang [$= 30$ chi];
upper width: [$c_1 =$] 2 zhang 4 chi [$= 24$ chi];
lower width: [$c_2 =$] 1 zhang 8 chi [$= 18$ chi];
length: [$l =$] 6 zhang 6 chi [$= 66$ chi].

County A:
　height: [$g_A =$] 1 zhang 5 chi [$= 15$ chi];
　length: [$l_A =$] 3 zhang 3 chi [$= 33$ chi];
　upper width: [$c_A =$] 2 zhang 1 chi [$= 21$ chi];

County B:
　height: [$g_B =$] 2 zhang 1 chi [$= 21$ chi];
　length: [$l_B =$] 1 zhang 3 chi 2 cun [$= 13.2$ chi];
　upper width: [$c_B =$] 2 zhang 2 chi 2 cun [$= 22.2$ chi].

County C:

> height: [g_C =] 3 *zhang* [= 30 *chi*];
> length: [l_C =] 1 *zhang* 9 *chi* 8 *cun* [= 19.8 *chi*];
> upper width: [c_C =] 2 *zhang* 4 *chi* [= 24 *chi*].

3.4.3. Method for calculating the width, length, and height of the dragon tail dyke

Multiply the work norm by the total number of men to obtain the volume [V] of the dyke.

$$V = (2{,}375 + 2{,}378 + 5{,}247) \text{ men} \times 1.98 \ chi^3/\text{man}$$
$$= 19{,}800 \ chi^3$$

求龍尾隄廣、袤、高術曰：以程功乘總人
為隄積，又六因之為虛積。以少高乘少袤
為隅冪，以少上廣乘之為鼈隅｛冪｜積｝。
以減虛積，餘，三約之，所得為實。并少
高、袤，以少上廣乘之，為鼈從橫廉冪。
三而一，加隅冪為方法。又三除少上廣，
以少袤、少高加之，為廉法，從。開立方
除之，得下廣。加差即高、廣、袤。

Multiply by 6 to obtain [K_1 =] the Space[1] Volume.

$$K_1 = 6V = 118{,}800 \ chi^3$$

Multiply the lessened height [$g{-}c_2$] by the lessened length [$l{-}c_2$] to obtain [K_2 =] the Area for the Corner.

$$K_2 = (g{-}c_2)(l{-}c_2) = 576 \ chi^2$$

Multiply this by the lessened upper width [$c_1{-}c_2$] to obtain [K_3 =] the Volume for the Corner *Bie*[*nao*].

$$K_3 = K_2(c_1{-}c_2) = (g{-}c_2)(l{-}c_2)(c_1{-}c_2) = 3{,}456 \ chi^3$$

Subtract this from the Space Volume [K_1] and divide the difference by 3. The result is the *shi* [the constant term of the equation].

$$shi = \frac{K_1 - K_3}{3} = \frac{6V - \left(g - c_2\right)\left(l - c_2\right)(c_1 - c2)}{3}$$

$$= 38{,}448 \ chi^3$$

Add together the lessened height and length [$g{-}c_2$, $l{-}c_2$] and multiply by the lessened upper width [$c_2{-}c_1$] to obtain [K_4 =] the Area for the *Bie*[*nao*] Crossing to the Side.

$$K_4 = \left[\left(g - c_2\right) + \left(l - c_2\right)\right]\left(c_1 - c_2\right) = 360 \ chi^2$$

---

1 *Xu* 虛. See fn. 1 in Section 2.1.3.3 above, p. 46.

Divide by 3 and add the Area for the Corner [K_2] to obtain the *fangfa* [the coefficient of the linear term].

$$fangfa = \frac{K_4}{3} + K_2 = 696 \ chi^2$$

Divide the lessened upper width [c_1-c_2] by 3. Add the lessened length [$l-c_2$] and lessened height [$g-c_2$] to obtain the *lianfa* [the coefficient of the quadratic term].

$$lianfa = \frac{c_1 - c_2}{3} + (l - c_2) + (g - c_2) = 62 \ chi$$

Extract the cube root to obtain the lower width [c_2].

$$c_2^{\ 3} + \left[\frac{c_1 - c_2}{3} + (l - c_2) + (g - c_2)\right]c_2^{\ 2} + \left(\frac{K_4}{3} + K_2\right)c_2$$

$$= \frac{K_1 - K_3}{3}$$

$$c_2^{\ 3} + 62c_2^{\ 2} + 696c_2 = 38,448 \ chi^3$$

$$c_2 = 18 \ chi$$

Add the differences to obtain the height [g], the [upper] width [c_1] and the length [l].

$$g = (g{-}c_2) + c_2 = 30 \ chi$$
$$c_1 = (c_1{-}c_2) + c_2 = 24 \ chi$$
$$l = (l{-}c_2) + c_2 = 60 \ chi$$

3.4.4. Method for calculating the volumes in cubic chi of the uniform contributions of the successive counties, and their width and length

◆ *3.4.4.1. The volumes of the contributions of the counties*

求逐縣均給積尺受廣、表，術曰：以程功乘當縣人為積尺。

Multiply the final amount of work by the [number of] men in the given county to obtain the volume in *chi*.

$$V_A = 2,375 \ men \times 1.98 \ chi^3/man = 4,702.5 \ chi^3$$
$$V_B = 2,378 \ men \times 1.98 \ chi^3/man = 4,708.44 \ chi^3$$
$$V_C = 5,247 \ men \times 1.98 \ chi^3/man = 10,389.06 \ chi^3$$

◆ *3.4.4.2. The dimensions of the contribution of County A*

For each [county], multiply the volume [$V_{A,B,C}$] by 6 and multiply by the area of [the square on] the length [l]. Let the divisor be the product of the height [g] and the difference in widths [c_1-c_2] and divide to make the *shi* [the constant term in the cubic equation].

$$shi_A = \frac{6V_A l^2}{(c_1 - c_2)g} = 682{,}803 \ chi^3 \qquad (3.11)$$

Triple the end width [c_2], multiply by the length [l], and divide by the difference between the widths [c_1-c_2] to obtain the *dulian* [the coefficient of the quadratic term].

$$dulian_A = \frac{3c_2 l}{c_1 - c_2} = 594 \ chi \qquad (3.12)$$

Extract the cube root. This is then [$l_A =$] the length [of the part built by] county A.

$$l_A{}^3 + \frac{3c_2 l}{c_1 - c_2} l_A{}^2 = \frac{6V_A l^2}{(c_1 - c_2)g}$$

$$l_A{}^3 + 594 l_A{}^2 = 682{,}803 \ chi^3$$

$$l_A = 33 \ chi$$

Multiply this by the full height [g] and divide by the full length [l]. This is [$g_A=$] the height [of the part built by] county A. Multiply the difference between the widths [c_1-c_2] by the length [of the part built by] county A [l_A], divide by the full length [l], and add the result to the end width [c_2]. This is [$c_A =$] the upper width [of the part built by] county A.

$$g_A = \frac{l_A g}{l} = 15 \ chi$$

$$c_A = \frac{l_A (c_1 - c_2)}{l} + c_2 = 21 \ chi$$

◆ *3.4.4.3. The dimensions of the contribution of County B*

The following calculation is the same as that given in Problem 3, Section 3.3.3.3 above.

各六因積尺。又乘袤冪,廣差乘高為法除之,為實。又三因末廣,以袤乘之,廣差而一,為都廉,從。開立方除之,即甲袤。以本高乘之,以本袤除之,即甲高。又以廣差乘甲袤,以本袤除之,所得,加末廣,即甲上廣。

其甲上廣即乙末廣，其甲高即垣高。求
{|實與}都廉，如前。又并甲上、下廣，
三之，乘甲高，以乘表冪，以法除之，
得垣方，從。開立方除之，即乙表。餘放
此。

The upper width [of the part built by] county A [c_A] is then the end width [of the part built by] county B and the height of [the part built by] county A [g_A] is then the height of the wall.

The calculation of [the *shi* and] the *dulian* [the constant term and the quadratic coefficient] is as before [compare (3.11) and (3.12)].

$$shi_B = \frac{6V_B\left(l - l_A\right)^2}{\left(c_1 - c_A\right)\left(g - g_A\right)} = 683{,}665.488 \; chi^3 \tag{3.13}$$

$$dulian_B = \frac{3c_A\left(l - l_A\right)}{c_1 - c_A} = 693 \; chi \tag{3.14}$$

Further add together the upper and lower widths of [the part built by] county A [c_A, c_2], triple this, and multiply by the height [of the part built by] county A [g_A]. Multiply by the area of [the square on] the length [$(l-l_A)^2$]. Divide by the divisor [see (3.13) above] to obtain the *yuanfang* 垣方 [the linear coefficient].

$$yuanfang_B = \frac{3g_A\left(c_A + c_2\right)\left(l - l_A\right)^2}{\left(c_1 - c_A\right)\left(g - g_A\right)} = 42{,}471 \; chi^2 \tag{3.15}$$

Extract the cube root. This is [$l_B =$] the length of [the part built by] county B.

$$\begin{aligned} l_B{}^3 + \frac{3c_A\left(l - l_A\right)}{c_1 - c_A} l_B{}^2 + \frac{3g_A\left(c_A + c_2\right)\left(l - l_A\right)^2}{\left(c_1 - c_A\right)\left(g - g_A\right)} l_B \\ = \frac{6V_B\left(l - l_A\right)^2}{\left(c_1 - c_A\right)\left(g - g_A\right)} \end{aligned} \tag{3.16}$$

$$l_B{}^3 + 693 l_B{}^2 + 42{,}471 l_B = 683{,}665.488 \; chi^3$$

$$l_B = 13.2 \; chi$$

The rest [of the calculations] resemble this.

◆ *3.4.4.4. Comment in smaller characters*

The following comment is obviously confused and/or corrupt. We have very tentatively interpreted it as fragments of a derivation of (3.16) above.

This 'dragon tail' resembles a *yanchu*. One *qiandu* and one *bie-nao* are put together side by side.

此龍尾猶羨除也。其塹堵一，鱉{腜|臑}
一，并而相連。今以表再乘積，廣差乘高
而一，所得，截鱉{腜|臑}表再自乘，為
立方一。{又各一鱉腜截表再自乘為立方
一|}又塹堵表自乘為冪{三|一}。又三因末
廣，以表乘之，廣差而一，與冪為高，故
為廉法。

See Figure 40, p. 100. The *qiandu* is *BDCTFE* and the *bienao* is *ATCF*.

Here the volume $[V_B]$ is multiplied twice by the length $[l–l_A]$ and divided by the difference between the widths $[c_1–c_A]$ multiplied by the height $[g–g_A]$.

$$\frac{V_B(l-l_A)^2}{(c_1-c_A)(g-g_A)}$$

Compare (3.4.3).

[. . .]

Something appears to be missing here.

The result, the length of the truncated *bienao* multiplied twice by itself $[l_B^3\,?]$, is one cube.

[. . .]

Again something appears to be missing.

The length of the *qiandu* $[l_B]$ is multiplied by itself to obtain one area $[l_B^2]$. The end width $[c_A]$, tripled, multiplied by the length $[l–l_A]$, and divided by the difference in widths $[c–c_A]$, is the height corresponding to the area $[l_B^2]$;

$$\frac{3c_A(l-l_A)}{c_1-c_A}$$

Compare (3.14). This result, multiplied by the 'area', l_B^2, is the quadratic term of (3.16).

this is why it is the *lianfa*.[1]

---

1 The quadratic coefficient in (3.16), called *dulian* in the main text.

3.5. Problem 5: A canal and a canal bank

3.5.1. The problem

In the following, consult Section 2.6.3 above and Figure 37, p. 94

假令穿河，表一里二百七十六步，下廣六步一尺二寸；北頭深一丈八尺六寸，上廣十二步二尺四寸；南頭深二百四十一尺八寸，上廣八十六步四尺八寸。運土於河西岸造潨，北頭高二百二十三尺二寸，南頭無高；下廣四百六尺七寸五釐，表與河同。甲郡二萬二千三百二十人，乙郡六萬八千七十六人，丙郡五萬九千九百八十五人，丁郡三萬七千九百四十四人。自穿、負、築，各人程功常積三尺七寸二分。限九十六日役河潨俱了。四郡分共造潨，其{[穿]}河自北頭先給甲郡，以次與乙{|、丙、丁}，合均賦積尺。問逐郡各給斜、正表，上廣及深，并潨上廣各多少？

Suppose a canal is to be excavated. The length is [$l =$] 1 *li* 276 *bu* [= 3,456 *chi*], the lower width is [$a =$] 6 *bu* 1 *chi* 2 *cun* [= 37.2 *chi*], the depth of the northern end is [$h_1 =$] 1 *zhang* 8 *chi* 6 *cun* [= 18.6 *chi*], and the upper width [of the northern end] is [$b_1 =$] 12 *bu* 2 *chi* 4 *cun* [= 74.4 *chi*]. The depth of the southern end is [$h_2 =$] 241 *chi* 8 *cun* [= 241.8 *chi*], and the upper width [of the southern end] is [$b_2 =$] 86 *bu* 4 *chi* 8 *cun* [= 520.8 *chi*].

The soil is moved to the canal's western shore to build a bank. The height of the northern end is [$g =$] 223 *chi* 2 *cun* [= 223.2 *chi*]. The southern end has no height. The lower width [at the northern end] is [$v_2 =$] 406 *chi* 7 *cun* 5 *li* [= 406.705 *chi*], and the length is the same as [the length of] the canal [l].

The volume of the canal, using the formula given in Section 3.3.3.5 above, Equation (3.10), is

$$V_{\text{canal}} = \frac{l}{6}\left[\left(2h_2 + h_1\right)\frac{a+b_2}{2} + \left(2h_1 + h_2\right)\frac{a+b_1}{2}\right]$$

$$= 89{,}672{,}832 \ chi^3$$

which is 4/3 of the volume of the bank, calculated in Section 3.5.3.3 below, Equation (3.23). This is because the volume of a tamped-earth structure is less than the volume of the original undisturbed earth; the ratio 4:3 for these quantities is given in *Jiuzhang suanshu*, Chapter 5, Problem 1.[1] Note also the calculation of V_A in Section 3.5.3.2 below, Equation (3.17).

In the following, consult Figure 38, p. 95.

Commandery A [sends] 22,320 men, commandery B [sends] 68,076 men, commandery C [sends] 59,985 men, commandery

---

1   Qian Baocong 1963: 159; Guo and Chemla 2004: 410–411; Guo Shuchun et al. 2013: 491–497.

D [sends] 37,944 men. [With regard to] the men digging, carrying, and ramming [the soil], the final amount of work for each man is a constant volume of 3 *chi* 7 *cun* 2 *fen* [i.e. 3.72 *chi*³] [of soil]. At the end of 96 days the work on the canal and the bank is completed.

The four commanderies share the construction of the bank. [The excavation of] the canal starts from the northern end. First [the work is] assigned to commandery A, thereafter in turn to B[, C, and D]. The equal assignment is in volumes of [cubic] *chi*.

Calculating each commandery's contribution [in terms of] slanted length [s_{A-D}], straight length [l_{A-D}], upper width [b_{2A-D}] and depth [h_{A-D}], together with the upper width of the bank [v_1], what is the measure of each?

3.5.2. Answer

Upper width of the bank, [v_1 =] 5 *zhang* 8 *chi* 2 *cun* 1 *fen* [= 58.21 *chi*].

Commandery A:
> straight length, [l_A =] 144 *zhang* [= 1,440 *chi*];
> slanted length, [s_A =] 144 *zhang* 3 *chi* [= 1,443 *chi*];
> upper width, [b_{2A} =] 26 *zhang* 4 *cun* [= 260.4 *chi*];
> depth, [h_A =] 11 *zhang* 1 *chi* 6 *cun* [= 111.6 *chi*].

Commandery B:
> straight length, [l_B =] 115 *zhang* 2 *chi* [= 1,152 *chi*];
> slanted length, [s_B =] 115 *zhang* 4 *chi* 4 *cun* [= 1,154.4 *chi*];
> upper width, [b_{2B} =] 40 *zhang* 9 *chi* 2 *cun* [= 409.2 *chi*];
> depth, [h_B =] 18 *zhang* 6 *chi* [= 186 *chi*].

Commandery C:
> straight length, [l_C =] 57 *zhang* 6 *chi* [= 576 *chi*];
> slanted length, [s_C =] 57 *zhang* 7 *chi* 2 *cun* [= 577.2 *chi*];
> upper width, [b_{2C} =] 48 *zhang* 3 *chi* 6 *cun* [= 483.6 *chi*];
> depth, [h_C =] 22 *zhang* 3 *chi* 2 *cun* [= 223.2 *chi*].

Commandery D:
> straight length, [l_D =] 28 *zhang* 8 *chi* [= 288 *chi*];
> slanted length, [s_D =] 28 *zhang* 8 *chi* 6 *cun* [= 288.6 *chi*];
> upper width, [b_{2D} =] 52 *zhang* 8 *cun* [= 528 *chi*];
> depth, [h_D =] 24 *zhang* 1 *chi* 8 *cun* [= 241.8 *chi*].

答曰：
滑上廣五丈八尺二寸一分。
甲郡正袤一百四十四丈，
　斜袤一百四十四丈三尺，
　上廣二十六丈四寸，
　深一十一丈一尺六寸。
乙郡正袤一百一十五丈二尺，
　斜袤一百一十五丈四尺四寸，
　上廣四十丈九尺二寸，
　深一十八丈六尺。
丙郡正袤五十七丈六尺，
　斜袤五十七丈七尺二寸，
　上廣四十八丈三尺六寸，
　深二十二丈三尺二寸。
丁郡正袤二十八丈八尺﹝六寸﹞，
　斜袤二十八丈八尺六寸，
　上廣五十二丈八寸，
　深二十四丈一尺八寸。

3.5.3. Method

術曰：如築隄術入之。

Use the method for 'Building a dyke' [in Problem 3, Section 3.3.3.2 above].

◆ 3.5.3.1. Comment in smaller characters

覆隄為河，彼注甚明，高深稍殊，程功是同，意可知也。

If one inverts the dyke [in Problem 3] to make the canal bed, that procedure is clear. The height and depth are slightly different, and the fixed amount of work is the same; the meaning is obvious.

The calculations in Sections 3.5.3.2–3.5.3.3 below are equivalent to those given in Problem 3, Section 3.3.3.3 above. Figures 37 and 38 use the same letter symbols as Figures 30 and 33 (pp. 82 and 87) in order to make clear the equivalence of the two problems.

◆ 3.5.3.2. The contributions of the commanderies

以程功乘甲郡人，又以{隄|限}日乘之，四之，三而一，為積。又六因，以乘表冪，以上廣差乘深差為法，除之，為實。又并小頭上、下廣，以乘小頭深，三之，為垣頭冪。又乘表冪，以法除之，為垣方。三因小頭上廣，以乘正表，以廣差除之，為都廉，從。開立方除之，即得小頭{|表|}，為甲表。求深、廣，以本表及深廣差求之{為法|}。以兩頭上廣差乘甲表，以本表除之，所得，加小頭上廣，即甲上廣。以小頭深減南頭深，餘，以乘甲表，以本表除之，所{|得}，加小頭深，即甲深。又正表自乘，深差自乘，并，而開方除之，即斜表。若求乙、丙、丁，每以前大深、廣為後小深、廣，準甲求之，即得。

Multiply the fixed amount of work by the number of men [sent by] commandery A. Multiply by the allotted number of days. Multiply this by 4, and divide by 3 to obtain the volume [V_A].

$$V_A = 3.72\ chi^3/\text{man·day} \times 22{,}320\text{ men}$$
$$\times\ 96\text{ days} \times 4/3 = 10{,}627{,}891.2\ chi^3 \quad (3.17)$$

Multiply by 6 and multiply by the area of [the square on] the length [l]. Multiply the difference between the upper widths [$b_2 - b_1$] by the difference between the depths [$h_2 - h_1$] to obtain [$K_2 =$] the Divisor. Divide to obtain the *shi* [the constant term].

$$K_2 = (b_2 - b_1)(h_2 - h_1) = 99{,}636.48\ chi$$

$$shi = \frac{6V_A l^2}{K_2} = 7{,}644{,}119{,}040\ chi^3 \quad (3.18)$$

Add the upper and lower widths of the small end [a, b_1] and multiply by the depth of the small end [h_1]. Triple this to obtain [$K_3 =$] the Area for the End of the Wall.

$$K_3 = 3(a + b_1)h_1 = 6{,}227.28\ chi^2 \quad (3.19)$$

Multiply by the area of [the square on] the length and divide by the Divisor [K_2] to make the *yuanfang* 垣方 [the linear coefficient].

$$yuanfang = \frac{K_3 l^2}{K_2} = 746,496 \; chi^2 \qquad (3.20)$$

Multiply the upper width of the small end [b_1] by 3 and multiply by the straight length. Divide by the difference between the widths to obtain the *dulian* [the coefficient of the quadratic term].

$$dulian = \frac{3b_1 l}{b_2 - b_1} = 1,728 \; chi \qquad (3.21)$$

Extract the cube root to obtain the small end, which is [$l_A =$] the length of [the part excavated by] A.

$$l_A{}^3 + \frac{3b_1 l}{b_2 - b_1} l_A{}^2 + \frac{K_3 l^2}{K_2} l_A = \frac{6 V_A l^2}{K_2} \qquad (3.22)$$

$$l_A{}^3 + 1728 l_A{}^2 + 746,496 l_A = 7,644,119,040 \; chi^3$$

$$l_A = 1,440 \; chi$$

Calculation of the depth and width: use the full length [l] and the differences between the [full] depths [$h_2 - h_1$] and widths [$b_2 - b_1$] to calculate them. Multiply the difference between the upper widths of the two ends [$b_2 - b_1$] by the length of [the part built by commandery] A [l_A] and divide by the full length [l]. Add the result to the upper width of the smaller end [b_1]. This is then the upper width of [the part excavated by] A.

$$b_{2A} = \frac{(b_2 - b_1) l_A}{l} + b_1 = 260.4 \; chi$$

Subtract the depth of the small end [h_1] from the depth of the southern end [h_2]. Multiply the difference by the length [of the part built by commandery] A [l_A] and divide by the full length [l]. Add the result to the depth of the small end [h_1]. This is then [$h_A =$] the depth [of the part built by] A.

$$h_A = \frac{(h_2 - h_1) l_A}{l} + h_1 = 111.6 \; chi$$

Multiply the straight length [l_A] by itself and multiply the difference between the depths [$h_A - h_1$] by itself. Add and extract the square root. This is [$s_A =$] the slanted length [of the part built by A].

$$s_A = \sqrt{l_A^2 + (h_A - h_1)^2} = 1{,}443 \; chi$$

In calculating [the dimensions of the parts built by commanderies] B, C and D, for each take the former larger depth and width to be the later smaller depth and width. Calculate in the same way as for [commandery] A; [the dimensions are] then obtained.

◆ 3.5.3.3. The width of the bank

求滑上廣術曰：以程功乘總人，又以限日乘之，為積。六因之，為實。以正袤除之，又以高除之，所得，以下廣減之，餘，又半之，即滑上廣。

The method for calculating the upper width of the bank: Multiply the fixed amount of work by the total number of men and multiply by the allotted number of days to obtain the volume [V].

$$V = 3.72 \; chi^3/\text{man·day} \times 188{,}325 \; \text{men} \times 96 \; \text{days}$$
$$= 67{,}254{,}624 \; chi^3 \tag{3.23}$$

Multiply this by six to obtain the dividend. Divide by the straight length [l] and divide by the height [g]. Subtract the lower width [v_2] from the result and halve the difference. This is then [$v_1 =$] the upper width of the bank.

$$v_1 = \frac{1}{2}\left(\frac{6V}{gl} - v_2\right) = 58.21 \; chi \tag{3.24}$$

3.6. Problem 6: A rectangular grain cellar

3.6.1. The problem

In the following, consult Section 2.5.2 above and Figure 29, p. 78.

Suppose four commanderies transport grain [and dig a cellar for it]. The norm for *hu* is 2 *chi* 尺 5 *cun* 寸 [i.e. 2.5 *chi³/hu*]. Each person's labour is taken to be the same. From the top [the work is] assigned to [commandery] A, then in turn to B, C, and D. Commandery A transports 38,745 *dan* 6 *dou* [= 38,745.6 *hu*], Commandery B transports 34,905 *dan* 6 *dou* [= 34,905.6 *hu*], Commandery C transports 26,270 *dan* 4 *dou* [= 26720.4 *hu*], and Commandery D transports 14,078 *dan* 4 *dou* [= 14,078.4 *hu*].

The four commanderies dig a cellar together. Its upper length is greater than the upper breadth by [$b_1 - a_1 =$] 1 *zhang* [= 10 *chi*], less than the lower length by [$b_2 - b_1 =$] 3 *zhang* [= 30 *chi*], greater than the depth by [$b_1 - h =$] 6 *zhang* [= 60 *chi*], and less than the lower breath by [$a_2 - b_1 =$] 1 *zhang* [= 10 *chi*]. Each commandery sends corvée labourers in proportion to the amount of grain transported. By the winter norm, for digging, carrying, and building, one person's labour is a volume of 12 [cubic] *chi* [per day]. The work is finished in one day.

What are the upper and lower breadth and length, and the depth, of the cellar? How many labourers are sent by each commandery, and what are the depth and breadth of each [part of the] cellar?

3.6.2. Answer

The cellar:

upper breadth, [$a_1 =$] 8 *zhang* [= 80 *chi*];
upper length [$b_1 =$] 9 *zhang* [= 90 *chi*];
lower breadth [$a_2 =$] 10 *zhang* [= 100 *chi*];
lower length [$b_2 =$] 12 *zhang* [= 120 *chi*];
depth [$h =$] 3 *zhang* [= 30 *chi*].

Commandery A, 8,072 persons;

depth [$h_A =$] 12 *chi*;
lower length [$b_A =$] 10 *zhang* 2 *chi* [= 102 *chi*];
breadth [$a_A =$] 8 *zhang* 8 *chi* [= 88 *chi*].

假令四郡輸粟，斛法二尺五寸。一人作功為均，自上給甲，以次與乙{|、丙、丁}。其甲郡輸粟三萬八千七百四十五石六斗，乙郡輸粟三萬四千九百五石六斗，丙郡輸粟二萬六千二百七十石四斗，丁郡輸粟一萬四千七十八石四斗。四郡共穿窖，上袤多於上廣一丈，少於下袤三丈，多於深六丈，少於下廣一丈。各計粟多少，均出丁夫。自穿、負、築，冬程人功常積一十二尺，一日役{|畢}。問窖上、下廣、袤、深，郡別出人及窖深、廣，各多少？

答曰：
窖上廣八丈，
上袤九丈，
下廣一十丈，
下袤一十二丈，
深三丈；
甲郡八千七十二人，
深一十二尺，
下袤一十丈二尺，
廣八丈八尺。
乙郡七千二百七十二人，
深九尺，
下袤一十一丈一尺，
廣九丈四尺。
丙郡五千四百七十三{尺|人}，
深六尺，
下袤一十一丈七尺，
廣九丈八尺。
丁郡二千九百三十三人，
深三尺，
下袤一十二丈，
廣一十丈。

Commandery B, 7,272 persons;
> depth [h_B =] 9 *chi*;
> lower length [b_B =] 11 *zhang* 1 *chi* [= 111 *chi*];
> breadth [a_B =] 9 *zhang* 4 *chi* [= 94 *chi*].

Commandery C, 5,473 persons;
> depth [h_C =] 6 *chi*;
> lower length [b_C =] 11 *zhang* 7 *chi* [= 117 *chi*];
> breadth [a_C =] 10 *zhang* [= 100 *chi*].

Commandery D, 2,933 persons;
> depth [h_D =] 3 *chi*;
> lower length [b_2 =] 12 *zhang* [= 120 *chi*];
> breadth [a_2 =] 10 *zhang* [= 100 *chi*].

3.6.3. The method for calculating the depth, breadth, and length

求窖深、廣、表術曰：以斜法乘總粟
為積尺，{|為實|¹}。又廣差乘表差，三而
一，為隅陽冪。乃置{塹|截}上廣，半廣
差加之，以乘{塹|截}上表，為隅頭冪。
又半表差乘{塹|截}上廣，以隅陽冪及隅
頭冪加之，為方法。又置{塹|截}上表及
{塹|截}上廣，并之為大廣。又并廣差及
表差，半之，以加大廣，為廉法，從。開
立方除之，即深。各加差，即合所問。

---

1 Guo Shuchun and Liu Dun 1998, 2: 11, 22 n. 55.

Multiply the norm for *hu* by the total amount of grain to make [V =] the volume in [cubic] *chi*, and take this as the *shi* [the constant term in the cubic equation].

$$shi = V = 114{,}000 \ hu \times 2.5 \ chi/hu = 285{,}000 \ chi^3$$

Further multiply the difference in breadths [a_2-a_1] by the difference in lengths [b_2-b_1] and divide by 3 to make [K_1 =] the Area for the Corner *Yang*[*ma*].

$$a_2 - a_1 = (a_2 - b_1) + (b_1 - a_1) = 20 \ chi$$

$$K_1 = \frac{(a_2 - a_1)(b_2 - b_1)}{3} = 200 \ chi^2$$

Then set up the truncated upper breadth [a_1-h], add to it one-half of the difference in breadths [a_2-a_1], and multiply by the truncated upper length [b_1-h] to make [K_2 =] the Area for the Corner Head.

$$a_1 - h = (b_1 - h) - (b_1 - a_1) = 50 \ chi$$

$$K_2 = \left[(a_1 - h) + \tfrac{1}{2}(a_2 - a_1) \right] (b_1 - h) = 3{,}600 \ chi^2$$

Further multiply one-half of the difference in lengths [b_2-b_1] by the truncated upper breadth [a_1-h]. Add to this the Area for the Corner *Yang*[*ma*] [K_1] and the Area for the Corner Head [K_2] to make the *fangfa* [the coefficient of the linear term].

$$fangfa = \tfrac{1}{2}(b_2 - b_1)(a_1 - h) + K_1 + K_2 = 4{,}550 \ chi^2$$

Further set up the truncated upper length $[b_1-h]$ and the truncated upper breadth $[a_1-h]$ and add to make $[K_3 =]$ the Large Breadth.

$$K_3 = (b_1-h) + (a_1-h) = 110 \ chi$$

Further add together the difference in breadths $[a_2-a_1]$ and the difference in lengths $[b_2-b_1]$, halve, and add the Large Breadth $[K_3]$ to make the *lianfa* [the coefficient of the quadratic term].

$$lianfa = \frac{(a_2 - a_1) + (b_2 - b_1)}{2} + K_3 = 135 \ chi$$

Extract the cube root; this is the depth.

$$h^3 + \left[\frac{(a_2 - a_1) + (b_2 - b_1)}{2} + K_3 \right] h^2$$
$$+ \left[\tfrac{1}{2}(b_2 - b_1)(a_1 - h) + K_1 + K_2 \right] h = V$$

$$h^3 + 135 \ h^2 + 4{,}550 \ h = 285{,}000 \ chi^3$$

$$h = 30 \ chi.$$

Add this to each of the differences; this corresponds to what is required.

$$b_1 = (b_1-h) + h = 90 \ chi$$
$$a_1 = (a_1-h) + h = (b_1-h) - (b_1-a_1) + h = 80 \ chi$$
$$b_2 = (b_2-h) + h = (b_1-h) + (b_2-b_1) + h = 120 \ chi$$
$$a_2 = (a_2-h) + h = (a_2-b_1) + (b_1-h) + h = 100 \ chi$$

3.6.4. The contributions of the commanderies

The method to calculate the volume in [cubic] *chi* assigned to each and the breadth, length, and depth that they receive is the same as the method for 'Building the [Grand Astrologer's] Platform' [Problem 2, Section 3.2.3.2 above].

Multiply the norm for *hu* by the amount of grain of Commandery A to make $[V_A =]$ the volume in [cubic] *chi*.

$$V_A = 38{,}745.6 \ hu \times 2.5 \ chi^3/hu = 96{,}864 \ chi^3$$

求均給積尺受廣、袤、深術曰：如築{隄|
臺}術入之。以斜法乘甲郡輸粟為積尺。
又三因，以深冪乘之，以廣差乘袤差而
一，為實。深乘上廣，廣差而一，為上
廣之高。深乘上袤，袤差而一，為上袤之
高。上廣之高乘上袤之高，三之，為方
法。又并兩高，三之，二而一，為廉法，
從。開立方除之，即甲深。以袤差乘之，
以本深除之，所得，加上袤，即甲下袤。
以廣差乘之，本深除之，所得，加{上}
廣，即甲下廣。若求乙、丙、丁，每以前
下廣、袤為後上廣、袤。以次，皆準此求
之，即得。若求人數，各以程功約當郡積
尺。

Further multiply by 3, multiply by the area of [the square on] the depth [h^2], and divide by the difference in breadths [$a_2 - a_1$] multiplied by the difference in lengths [$b_2 - b_1$] to make the *shi* [the constant term in the equation].

$$shi = \frac{3V_A h^2}{(a_2 - a_1)(b_2 - b_1)} = 435,888 \ chi^3$$

The depth multiplied by the upper breadth [ha_1], divided by the difference in breadths [$a_2 - a_1$], is [$K_4 =$] the Height for the Upper Breadth.

$$K_4 = \frac{ha_1}{(a_2 - a_1)} = 120 \ chi$$

The depth multiplied by the upper length [db_1], divided by the difference in lengths [$b_2 - b_1$], is [$K_5 =$] the Height for the Upper Length.

$$K_5 = \frac{hb_1}{(b_2 - b_1)} = 90 \ chi$$

The Height for the Upper Breadth [K_4], multiplied by the Height for the Upper Length [K_5] and multiplied by 3, is the *fangfa* [the coefficient of the linear term].

$$fangfa = 3K_4 K_5 = 32,400 \ chi^2$$

Further add together the two Heights [K_4 and K_5], multiply by 3, and divide by 2 to make the *lianfa* [the coefficient of the quadratic term].

$$lianfa = \frac{3(K_4 + K_5)}{2} = 315 \ chi$$

Extract the cube root; this is the depth of [the contribution of Commandery] A.

$$h_A{}^3 + \frac{3(K_4 + K_5)}{2} h_A{}^2 + 3K_4 K_5 h_A = \frac{3V_A h^2}{(a_2 - a_1)(b_2 - b_1)}$$

$$h_A{}^3 + 315 \ h_A{}^2 + 32,400 \ h_A = 435,888 \ chi^3$$

$$h_A = 12 \ chi.$$

Multiply this by the difference in lengths $[b_2 - b_1]$, divide it by the full depth $[h]$, and add to the result the upper length $[b_1]$; this is the lower length of [the contribution of Commandery] A.

$$b_A = \frac{h_A (b_2 - b_1)}{h} + b_1 = 102 \ chi$$

Multiply it $[h_A]$ by the difference in breadths $[a_2 - a_1]$, divide by the full depth $[h]$, and add the upper breadth $[a_1]$; this is $[a_A =]$ the upper breadth of [the contribution of Commandery] A.

$$a_A = \frac{h_A (a_2 - a_1)}{h} + a_1 = 88 \ chi$$

If [the contributions of Commanderies] B, C, and D are calculated, in each case the previous lower breadth and length are taken as the later upper breadth and height. Then this calculation is taken as the standard, and they are obtained.

If the numbers of persons are calculated, in each case the labour norm is divided into the respective commandery's volume in [cubic] *chi*.

3.7. Problem 7: A square granary

3.7.1. The problem

In the following, consult Section 2.1.1 above and Figure 8, p. 35.

假令亭倉，上小、下大。上下方差六尺，高多上方九尺，容粟一百八十七石二斗。今已運出五十石四斗。問倉上、下方、高及餘粟深、上方各多少？

Suppose there is a granary [in the form of a] [*fang*]*ting*. The top is small and the bottom is large. The difference between [the sides of] the upper and lower squares is [*a–b* =] 6 *chi*. The height is [*h–b* =] 9 *chi* larger than [the side of] the upper square. [The granary] contains 187 *dan* 2 *dou* [= 187.2 *hu*] of grain. 50 *dan* 4 *dou* [= 50.4 *hu*] are removed. What are [the sides of] the upper and lower squares [*b*, *a*], the height [*h*] of the granary, and the depth [*g*] and [the side of] the upper square [*c*] of the remaining grain?

3.7.2. Answer

答曰：
　上方三尺，
　下方九尺，
　高一丈二尺。
　餘粟深、上方俱六尺。

[Side of the] upper square, [*b* =] 3 *chi*;
[side of the] lower square, [*a* =] 9 *chi*;
height, [*h* =] 1 *zhang* 2 *chi* [= 12 *chi*];
depth and [side of the] upper square of the remaining grain each
　　[*g* = *c* =] 6 *chi*.

3.7.3. Method

◆ *3.7.3.1. The dimensions of the granary*

求倉方、高術曰：以斜法乘容粟為積尺。又方差自乘，三而一，為隅陽冪。以乘截高，以減積，餘為實。又方差乘截高，加隅陽冪，為方法。又置方差，加截高，為廉法，從。開立方除之，即上方。加差，即合所問。

The method for calculating [the sides of] the squares and height of the granary is: Multiply the norm for *hu* [2.5 *chi*^{3}/*hu*] by the [amount of] grain contained to obtain the volume [*V*] in [cubic] *chi*.

$$V = 187.2 \ hu \times 2.5 \ chi^3/hu = 468 \ chi^3$$

Multiply the difference between [the sides of] the squares [*a–b*] by itself and divide by 3 to obtain [*K*$_1$ =] the Area for the Corner *Yang*[*ma*].

$$K_1 = \frac{(a-b)^2}{3} = 12 \ chi^2$$

Multiply by the truncated height [$h-b$] and subtract [the product] from the volume [V]. The difference is the *shi* [the constant term of the cubic equation].

$$shi = V - K_1 (h-b) = 360 \ chi^3$$

Multiply the difference between [the sides of] the squares [$a-b$] by the truncated height [$h-b$] and add the Area for the Corner [*Yang*]*ma* [K_1] to obtain the *fangfa* [the coefficient of the linear term].

$$fangfa = (a-b)(h-b) + K_1 = 66 \ chi^3$$

Lay out the difference between [the sides of] the squares [$a-b$] and add the truncated height [$h-b$] to obtain the *lianfa* [the coefficient of the quadratic term].

$$lianfa = (a-b) + (h-b) = 15 \ chi$$

Extract the cube root. This is [the side of] the upper square [b].

$$b^3 + \left[(a-b)+(h-b)\right]b^2$$
$$+ \left[(a-b)(h-b)+K_1\right]b = V - K_1(h-b)$$

$$b^3 + 15b^2 + 66b = 360 \ chi^3$$

$$b = 3 \ chi.$$

Add the differences; [the results] correspond to what was asked.

$$a = b + (a-b) = 9 \ chi$$
$$h = b + (h-b) = 12 \ chi$$

◆ *3.7.3.2. The dimensions of the remaining grain*

• *3.7.3.2.1. The constant term of the equation*

The method for calculating the height and [the side of] the upper square of the remaining grain is: Multiply the norm for *hu* [2.5 *chi³* / *hu*] by the amount of grain removed.

求餘粟高及上方術曰：以斛法乘出粟，三之，以乘高冪，令方差冪而一，為實。

$$W = 50.4 \ hu \times 2.5 \ chi^3 \ / \ hu = 126 \ chi^3$$

Triple this and multiply by the area of [the square on] the height [h]. Divide by the area of [the square on] the difference between [the sides of] the squares [$a-b$] to obtain the *shi* [the constant term of the cubic equation].

$$shi = \frac{3Wh^2}{(a-b)^2} = 1,512 \ chi^3 \tag{3.25}$$

• *3.7.3.2.2. Comment in smaller characters*

此是大、小高各自乘又相乘，各乘取
高。{是 | 凡}大高者，即是取高與小高
并。

This is the products of the Large and Small Heights [K_3 and K_2] each by itself and by each other and each multiplied by the height of what was removed [g]. The Large Height is the sum of the height of what was removed [g] and the Small Height [K_2].

The Small Height, K_2, will be defined in the main text, directly below (see (3.28)).

$$K_3 = K_2 + g \tag{3.26}$$

The three products mentioned in the first sentence are

$$gK_3{}^2, gK_2{}^2, gK_2K_3 \tag{3.27}$$

• *3.7.3.2.3. The rest of the equation*

高乘上方，方差而一，為小高。令自乘，
三之，為方法。三因小高為廉法，從。開
立方除之，得取出高。以減本高，餘即殘
粟高。置出粟高，又以方差乘之，以本高
除之，所得，加上方，即餘粟上方。

Multiply the height [h] by [the side of] the upper square [b] and divide by the difference between [the sides of] the squares [$a-b$] to obtain [$K_2 =$] the Small Height.

$$K_2 = \frac{hb}{a-b} = 6 \ chi \tag{3.28}$$

Multiply this by itself and triple it to obtain the *fangfa* [the coefficient of the linear term].

$$fangfa = 3K_2{}^2 = 108 \ chi^2$$

Multiply the Small Height [K_2] by 3 to make the *lianfa* [the coefficient of the quadratic term].

$$lianfa = 3K_2 = 18 \ chi$$

Extract the cube root to obtain the height of the removed [grain].

$$g^3 + 3K_2 g^2 + 3K_2{}^2 g = \frac{3Wh^2}{(a-b)^2}$$

$$g^3 + 18g^2 + 108g = 1{,}512\ chi^3$$

$$g = 6\ chi.$$

Subtract [this] from the full height [h]; the difference is then the height of the remaining grain [$h{-}g = 6\ chi$].

Lay out the height of the removed grain [g] and multiply by the difference between [the sides of] the squares [$a{-}b$]. Divide by the full height [h] and add the result to [the side of] the upper square [b]. This is then [the side of] the upper square of the surplus grain [c].

$$c = \frac{g(a-b)}{h} + b = 6\ chi$$

• *3.7.3.2.4. Comment in small characters*

Here the original method is: [The sides of] the upper and lower squares [b, c] are multiplied by each other, and each is multiplied by itself. The sum is multiplied by the height [g] and divided by 3.

$$W = \frac{(bc + b^2 + c^2)g}{3}$$

To return to the starting point, this is tripled, then multiplied by the area of [the square on] the height [h] and divided by the area of [the square on] the difference [$a{-}b$]. The result is the number obtained by 'multiplying the Large Height [K_3] by the Small Height [K_2] and multiplying each by itself' [and multiplying each by the height of the removed grain, g, as stated in Section 3.7.3.2.2 above].

$$\frac{3Wh^2}{(a-b)^2} = gK_2 K_3 + gK_2{}^2 + gK_3{}^2$$

Why is this? Multiplying the height [h] by [the side of] the lower square [c] and dividing by the difference between [the sides of] the squares [a, b] gives the Large Height [K_3].

此本術曰：上下方相乘，又各自乘，并以高乘之，三而一。今還元，三之，又高冪乘之，差冪而一，得大小高相乘，又各自乘之數。何者？若高乘下方，方差而一，得大高也。若高乘上方，方差而一，得小高也。然則斯本下方自乘，故須高{|冪}乘之，差自乘而一，即得大高自乘之數。小高亦然。凡大高者，即是取高與小高并，相連。今大高自{陳|乘}為大方。大方之內即有取高自乘冪一，隅頭小高自乘冪一。又其兩邊各{一|有}以取高乘小高為冪二。又大小高相乘為中方。中方之內即有小高乘取高冪一，又小高自乘即是小方之冪又一。則小高乘大高，又各自乘三等冪，皆以乘取高為立積。故三因小冪為方，及三小高為廉也。

$$K_3 = g + K_2 = \frac{c-b}{a-b}h + \frac{b}{a-b}h = \frac{ch}{a-b}$$

Multiplying the height [h] by [the side of] the upper square [b] and dividing by the difference between [the sides of] the squares [a, b] gives the Small Height [K_2].

This restates (3.28).

However, in this case the original [side of the] lower square [b] is multiplied by itself; therefore it is necessary to multiply by the area of [the square on] the height [h] and divide by the difference [$a-b$] multiplied by itself. Then the number of the Large Height [K_3] multiplied by itself is obtained. It is the same with the Small Height [K_2].

$$\frac{c^2 h^2}{(a-b)^2} = \left(\frac{ch}{a-b}\right)^2 = K_3^{\ 2}$$

$$\frac{b^2 h^2}{(a-b)^2} = \left(\frac{bh}{a-b}\right)^2 = K_2^{\ 2}$$

In the following, consult Figure 10, p. 42.

The Large Height [K_3] is the height of the removed [grain] [g] and the Small Height [K_2] added and linked together [see (3.26)]. The product of the Large Height [K_3] by itself is a large square [$ABCD$]. Inside the large square are: one area [$DFGJ$] which is the product of the height of what was removed [g] by itself; one area [$BHGE$] at the corner which is the product of the Small Height [K_2] by itself; and two areas [$AEGF$ and $CJGH$] at the two sides which are [each] the product of the height of what was removed [g] by the Small Height [K_2].

$$K_3^{\ 2} = (K_2 + g)^2 = g^2 + K_2^{\ 2} + 2gK_2$$

Further, the product of the Large Height [K_3] by the Small Height [K_2] is the middle rectangle [$BCJE$]. Inside the middle rectangle are one area [$CJGH$] which is the product of the Small Height [K_2] by the height of what was removed [g] and the area of the small square [$BHGE$] which is the product of the Small Height [K_2] by itself.

$$K_2 K_3 = K_2(K_2 + g) = gK_2 + K_2^{\ 2}$$

Then the three areas given by the product of the Large and Small Heights [K_3, K_2] and each of their products with themselves are each multiplied by the height of what was removed [g] to make volumes.

The three volumes are those listed in (3.27).

Therefore three times the area of [the square on] the Small [Height] [K_2] is the *fangfa* [the coefficient of the linear term] and three times the Small Height [K_2] is the *lianfa* [the coefficient of the quadratic term].

3.8. Problem 8: A wedge

3.8.1. The problem

In the following, consult Section 2.4 above and Figure 22, p. 60.

假令芻甍，上袤三丈，下袤九丈，廣六丈，高一十二丈。有甲縣六百三十二人，乙縣二百四十三人。夏程人功{當|常}積三十六尺，限八日役{||畢¹}，自穿築。二縣共造。今甲縣先到。問自下給高、廣、袤各多少？

---

1 Guo Shuchun and Liu Dun 1998, 2: 13, 22 n. 64.

Suppose a *chumeng* [is to be constructed of tamped earth]. The upper length is [l_1 =] 3 *zhang* [= 30 *chi*], the lower length is [l_2 =] 9 *zhang* [= 90 *chi*], the breadth is [b =] 6 *zhang* [= 60 *chi*], and the height is 12 *zhang* [= 120 *chi*].

County A has 632 persons and County B has 243 persons. By the summer norm one person's task is a constant volume of 36 [cubic] *chi* [of earth]. In eight days the corvée labour is finished. Between them, the counties build [the *chumeng*], each both digging and tamping [earth].

County A arrives first. What are the height, breadth, and length [h_A, b_B, l_B] of its assignment, starting from the bottom?

3.8.2. Answer

答曰：
　高四丈八尺，
　上廣三丈六尺，
　袤六丈六尺。

Height [h_A =] 4 *zhang* 8 *chi* [= 48 *chi*];
upper breadth [b_B =] 3 *zhang* 6 *chi* [= 36 *chi*];
length [l_B =] 6 *zhang* 6 *chi* [= 66 *chi*].

3.8.3. Method

求甲縣均給積尺，受廣、袤術曰：以程功乘乙縣人數，又以限日乘之，為積尺。以六因之，又高冪乘之，又袤差乘廣而一，所得，又半之為實。高乘上袤，袤差而一，為上袤之高。三因上袤之高，半之，為廉法，從。開立方除之，得乙高。以減甍高，餘即甲高。求廣、袤，依率求之。

The method to calculate the volume in [cubic] *chi*, the breadth, and the length of the assignment of County A [V_A, b_B, l_B] is: Multiply the number of persons of County B by the labour norm and further multiply this by the number of days to make [V_B =] the volume in [cubic] *chi*.

$$V_B = 243 \text{ persons} \times 36 \ chi^3/\text{person·day} \times 8 \text{ days}$$
$$= 69{,}984 \ chi^3$$

Multiply this by 6, further multiply by the area of [the square on] the height [h^2], and further divide by the difference in lengths [$l_2 - l_1$] multiplied by the breadth [b]. Further halve the result to make the *shi* [the constant term of the equation].

– 174 –

$$shi = \frac{6V_{\mathrm{B}}h^2}{2(l_2 - l_1)b} = 839{,}808 \ chi^3$$

The height [h] multiplied by the upper length [l_1], divided by the difference in lengths [$l_2 - l_1$], is [$K_1 =$] the Height for the Upper Length.

$$K_1 = \frac{hl_1}{l_2 - l_1} = 60 \ chi$$

Multiply the Height for the Upper Length [K_1] by 3 and halve this to make the *lianfa* [the coefficient of the quadratic term].

$$lianfa = \frac{3K_1}{2} = 90 \ chi$$

Extract the cube root to obtain the height [h_{B}] of [the assignment of County] B.

$$h_{\mathrm{B}}{}^3 + \frac{3K_1}{2}h_{\mathrm{B}}{}^2 = \frac{6V_{\mathrm{B}}h^2}{2(l_2 - l_1)b}$$

$$h_{\mathrm{B}}{}^3 + 90h_{\mathrm{B}}{}^2 = 839{,}808 \ chi^3$$
$$h_{\mathrm{B}} = 72 \ chi$$

Subtract this from the height of the [*chu*]*meng* [h]; the difference is [$h_{\mathrm{A}} =$] the height of [the assignment of County] A.

$$h_{\mathrm{A}} = h - h_{\mathrm{B}} = 48 \ chi$$

Calculate the breadth and length [$b_{\mathrm{B}}, l_{\mathrm{B}}$] using proportions.

$$b_{\mathrm{B}} = \frac{h_{\mathrm{B}}}{h}b = 36 \ chi$$

$$l_{\mathrm{B}} = \frac{h_{\mathrm{B}}}{h}(l_2 - l_1) + l_1 = 66 \ chi$$

3.8.4. Comment in smaller characters

This 'B volume' [V_{B}] was originally [obtained by] doubling the lower length [l_{B}], adding the upper length [l_1], multiplying by the lower breadth [b_{B}] and the height [h_{B}], and dividing by 6; this makes the volume of a [*chu*]*meng*.

此乙積本倍下表，上表從之。以下廣及高乘之，六而一，為一嚢積。今還元須六因之，以高冪乘之為實。｛乘｜｝表差乘廣而一。得取高自乘以乘｛二上表之高｜三上表之高｜上表之高者三¹｝，｛并大廣表相連之數｜｝則三小高為廉法，各以取高為方。仍有取高為立方者｛｜二｝。故半之，為立方一，又須半廉法。

1 Li Huang 1832, *xia* 下, 30b.

$$V_{\mathrm{B}} = \frac{(2l_{\mathrm{B}} + l_1)b_{\mathrm{B}}h_{\mathrm{B}}}{6}$$

Here, returning to the origin, this is multiplied by 6 and by the area of [the square on] the height [h^2] (to make the *shi*)[1] and divided by the difference in lengths [$l_2 - l_1$] multiplied by the breadth [b]. The result is the assigned height [*qu gao* 取高] multiplied by itself [h_{B}^2] multiplied by 3 times the Height for the Upper Length [K_1]. Therefore three of the smaller Height [i.e., the Height for the Upper Length, K_1] is the *lianfa* (in each case taking the smaller Height as a square).[2] There remain two cubes made of the assigned height [h_{B}]; therefore it is halved to make one cube, and the *lianfa* must also be halved.

$$\frac{6V_{\mathrm{B}}h^2}{2(l_2 - l_1)b} = h_{\mathrm{B}}^3 + \frac{3K_1}{2}h_{\mathrm{B}}^2$$

---

1 This phrase appears to be out of place.

2 The function of this clause is unclear.

3.9. Problem 9: A circular granary

3.9.1. The problem

In the following, consult Section 2.1.2 and Figures 11 and 12, p. 43.

Suppose there is a circular granary which is smaller at the top and larger at the bottom. The norm for *hu* is 2 *chi* 5 *cun* [i.e. 2.5 *chi³*/ *hu*]. The proportions to be used are diameter 1 circumference 3 [i.e., the approximation $\pi \approx 3$ is used]. The difference between the top and bottom circumferences is [$c_2 - c_1 =$] 1 *zhang* 2 *chi* [= 12 *chi*], and the height is greater than the upper circumference by [$h - c_1 =$] 1 *zhang* 8 *chi* [= 18 *chi*]. It holds 705 *hu* 6 *dou* [= 705.6 *hu*] of grain.

Now 266 *dan* 4 *dou* [= 266.4 *hu*] of grain has been removed. What are the distance from the remaining grain to the top [h_1], the upper and lower circumferences [c_1, c_2], and the height [h]?

假令圓囷上小、下大。斜法二尺五寸。以率徑一周三。上下周差一丈二尺，高多上周一丈八尺。容粟七百五斛六斗。今已運出二百六十六石四斗。問殘粟去口上、下周、高，各多少？

3.9.2. Answer

Upper circumference, [$c_1 =$] 1 *zhang* 8 *chi* [= 18 *chi*];
lower circumference, [$c_2 =$] 3 *zhang* [= 30 *chi*];
height, [$h =$] 3 *zhang* 6 *chi* [= 36 *chi*];
distance from top, [$h_1 =$] 1 *zhang* 8 *chi* [= 18 *chi*];
circumference of grain, [$c_3 =$] 2 *zhang* 4 *chi* [= 24 *chi*].

答曰：
上周一丈八尺，
下周三丈，
高三丈六尺，
去口一丈八尺，
粟周二丈四尺。

3.9.3. The dimensions of the granary

The method to calculate the upper and lower circumferences and the height of the granary is: Multiply the norm for *hu* by the amount of grain it holds; further multiply by 36 and divide by 3 to obtain [$F=$] the volume of a *fangting*.

$$V = 2.5 \ chi^3/hu \times 705.6 \ hu = 1{,}764 \ chi^3$$

$$F = \frac{36 \ V}{3} = 21{,}168 \ chi^3$$

Further multiply the difference in circumferences [$c_2 - c_1$] by itself and divide by 3 to make [$K_1 =$] the Area for the Corner *Yang*[*ma*].

求圓囷上、下周及高術曰：以斜法乘容粟，又三十六乘之，三而一，為方亭之積。又以周差自乘，三而一，為隅陽冪。以乘截高，以減亭積，餘為實。又周差乘截高，加隅陽冪，為方法。又以周差加截高為廉法，從。開立方除之，得上周。加差而合所問。

– 177 –

$$K_1 = \frac{(c_2 - c_1)^2}{3} = 48 \ chi^2$$

Multiply this by the truncated height $[h{-}c_1]$ and subtract from the volume of the *fangting* [*F*]; the remainder is the *shi* [the constant term in the equation].

$$shi = F - K_1(h - c_1) = 20{,}304 \ chi^3$$

Further multiply the difference $[c_2{-}c_1]$ by the truncated height $[h{-}c_1]$ and add the Area for the Corner *Yang*[*ma*] [K_1] to make the *fangfa* [the coefficient of the linear term].

$$fangfa = (c_2 - c_1)(h - c_1) + K_1 = 264 \ chi^2$$

Further add the difference in circumferences $[c_2{-}c_1]$ and the truncated height $[h{-}c_1]$ to make the *lianfa* [the coefficient of the quadratic term].

$$lianfa = (c_2 - c_1) + (h - c_1) = 30 \ chi$$

Extract the cube root to obtain the upper circumference [c_1].

$$c_1^3 + \left[(c_2 - c_1) + (h - c_1)\right]c_1^2 + \left[(c_2 - c_1)(h - c_1) + K_1\right]c_1$$
$$= F - K_1(h - c_1)$$
$$c_1^3 + 30c_1^2 + 264c_1 = 20{,}304 \ chi^3$$

$$c_1 = 18 \ chi$$

Adding the differences gives what was asked for.

$$c_2 = c_1 + (c_2{-}c_1) = 30 \ chi$$
$$h = c_1 + (h{-}c_1) = 36 \ chi$$

3.9.4. The grain

求粟去口術曰：以斜法乘出斜，三十六乘之，以乘高冪，如周差冪而一，為實。高乘上周，周差而一，為小高。今自乘，三之為方法。三因小高為廉法，從。開立方除之，即去口。

The method to calculate the distance of the grain from the mouth [of the granary] is: Multiply the norm for *hu* by the amount of grain removed, multiply by 36, multiply by the area of [the square on] the height [h^2], and divide by the area of [the square on] the difference between the circumferences $[(c_2-c_1)^2]$ to make the *shi* [the constant term].

$$W = 2.5\ chi^3/hu \times 266.4\ hu = 666\ chi^3$$

$$shi = \frac{36Wh^2}{(c_2 - c_1)^2} = 215,784\ chi^3$$

The height $[h]$ multiplied by the upper circumference $[c_1]$, divided by the difference of the circumferences $[(c_2-c_1)]$ makes $[K_2 =]$ the Small Height.

$$K_2 = \frac{hc_1}{c_2 - c_1} = 54\ chi$$

Multiply this by itself and multiply by 3 to make the *fangfa* [the coefficient of the linear term].

$$fangfa = 3K_2^2 = 8,748\ chi^2$$

Multiply the Small Height $[K_2]$ by 3 to make the *lianfa* [the coefficient of the quadratic term].

$$lianfa = 3K_2 = 162\ chi$$

Extract the cube root; this is $[h_1 =]$ the distance from the mouth.

$$h_1^3 + 3K_2h_1^2 + 3K_2^2h_1 = \frac{36Wh^2}{(c_2 - c_1)^2}$$

$$h_1^3 + 162\ h_1^2 + 8,748\ h_1 = 215,784\ chi^3$$

$$h_1 = 18\ chi$$

3.9.5. Comment in smaller characters

Multiplication by 36 gives a truncated *fangting*; it is not different from the previous rectangular cellar [Problem 7, Section 3.7.3.2 above].[1]

三十六乘訖，即是截方亭，{之|與I之¹}前方窖不別。

---

[1] Guo Shuchun and Liu Dun 1998, 2: 14, 23, n. 68.

---

[1] The sentence seems grammatically later than the rest of the text. Both *qi* 訖 as an aspect marker and *shi* 是 as a copula are rare before the Song period.

3.9.6. The remaining dimension

置去口，以周差乘之，以本高除之，所{|
得}，加上周，即粟周。

Lay out the distance from the mouth [h_1], multiply it by the difference of the circumferences [$c_2 - c_1$], and divide it by the full height [h]. Add the result to the upper circumference [c_1]; this is [$c_3 =$] the circumference of the grain.

$$c_3 = \frac{h_1(c_2 - c_1)}{h} + c_1 = 24 \; chi$$

3.10. Problem 10: A square granary and a circular storage pit

3.10.1. The problem

In the following, consult Section 2.3.1 above and Figure 16, p. 53.

Suppose there is grain, 23,120 *hu* 7 *dou* 3 *sheng* [= 23,120.73 *hu*].

It is desired to make one square granary and one circular storage pit so that the grain exactly fills both. Let the height [*h*] and the depth be equal, the side of the square [*s*] be less than the diameter of the circle [*d*] by [*d–s* =] 9 *cun* [= 0.9 *chi*], and greater than the height by [*s–h* =] 2 *zhang* 9 *chi* 8 *cun* [= 29.8 *chi*]. The proportions are to be diameter 7 circumference 22 [i.e. use the approximation $\pi \approx {}^{22}/_7$]. What are the [side of the] square [*s*], the diameter [*d*], and the depth [*h*]?

假令有粟二萬三千一百二十斛七斗三升。欲作方倉一，圓窖一，盛各滿中而粟適盡。令高深等，使方面少於圓徑九寸，多於高二丈九尺八寸。率徑七、周二十二。問方、徑、深多少？

3.10.2. Answer

The [side of the] square of the granary, [*s* =] 4 *zhang* 5 *chi* 3 *cun* [= 45.3 *chi*]. (It holds 12,722 *hu* 9 *dou* 5 *sheng* 8 *ge* [= 12,722.958 *hu*].)[1]

The diameter of the storage pit, [*d* =] 4 *zhang* 6 *chi* 2 *cun* [= 46.2 *chi*]. (It holds 10,397 *dan* 7 *dou* 7 *sheng* 2 *ge* [=10,397.772 *hu*].)[2]

The height and depth, each [*h* =] 1 *zhang* 5 *chi* 5 *cun* [= 15.5 *chi*].

答曰：
　倉方四丈五尺三寸
　（容粟一萬二千七百二十二斛九斗五升八合。）
　窖徑四丈六尺二寸
　（容粟一萬三百九十七石七斗七升二合。）
　高與深各一丈五尺五寸。

3.10.3. Method

The method to calculate the [side of the] square, the diameter, the height, and the depth: Multiply the norm for *hu* by 14, multiply by the amount of grain, and divide by 25 to make the *shi* [the constant term in the equation].

$$V = 2.5 \ chi^3/hu \times 23,120.73 \ hu = 57,801.825 \ chi^3$$

求方徑高深術曰：十四乘斛法，以乘粟數，二十五而一，為實。又倍多加少，以乘少數，又十一乘之，二十五而一。多自乘加之，為方法。又倍少數，十一乘之，二十五而一，又倍多加之，為廉法，從。開立方除之，即高、深。各加差即方徑。

---

1 Comment in smaller characters.

2 Comment in smaller characters.

$$shi = \frac{14}{25}V = 32{,}369.022 \ chi^3$$

Further double the 'greater than' [s–h], add the 'less than' [d–s], and multiply this by the amount 'less than' [d–s]. Further multiply by 11, divide by 25, and add the 'greater than' multiplied by itself [$(s-h)^2$] to make the *fangfa* [the coefficient of the linear term].

$$fangfa = \frac{11}{25}\big[2(s-h)+(d-s)\big](d-s)+(s-h)^2$$

$$= 911.998 \ chi^2 \qquad\qquad (3.29)$$

Further double the amount 'less than' [d–s], multiply this by 11, and divide by 25. Further add twice the 'greater than' [s–h] to make the *lianfa* [the coefficient of the quadratic term].

$$lianfa = 2(d-s)\frac{11}{25}+2(s-h) = 60.392 \ chi \qquad (3.30)$$

Extract the cube root; this is the height and the depth [h].

$$h^3 + \left[2(d-s)\frac{11}{25}+2(s-h)\right]h^2$$
$$+ \left[\frac{11}{25}\big[2(s-h)+(d-s)\big](d-s)+(s-h)^2\right]h = \frac{14}{25}V$$

$$h^3 + 60.392 \ h^2 + 911.998 \ h = 32{,}369.022 \ chi^3$$
$$h = 15.5 \ chi$$

Adding each of the differences gives [the side of] the square [s] and the diameter [d].

$$s = (s\!-\!h) + h = 45.3 \ chi$$
$$d = (d\!-\!s) + s = 46.2 \ chi$$

3.10.4. Comment in smaller characters

◆ *3.10.4.1. An alternative calculation of the quadratic coefficient, fangfa*

The norm for *hu* is multiplied by 14 and multiplied by the [amount of] grain to make a volume in [cubic] *chi*. There was an earlier division by 14; now, returning to the origin, there is a multiplication

一十四乘斜法以乘粟為積尺。前一十四
除，今還元一十四乘。為徑自乘者{是‖
是¹}一十一，方自乘者{是‖是²}一十四，
故并之為二十五。凡此方圓二徑長短不
同，二徑各自乘為方，大小各別。然則此
{壍丨截}方二丈九尺八寸，{壍丨截}徑三
丈七寸皆成{立方丨方面}。此應{壍丨截}方
自乘一十四乘之，{壍丨壍丨截³}徑自乘一
十一乘之，二十五而一，為隅幂，即方法
也。但二隅{方丨幂丨方⁴}皆以{壍丨截}數為
方面。今此術就省，

1 Guo Shuchun and Liu Dun 1998, 2: 14, 23 n. 70.
2 Ibid.
3 Ibid. n. 56.
4 Ibid. n. 74.

by 14. This is 11 of the diameter multiplied by itself [d^2] and 14 of the [side of the] square multiplied by itself [s^2].

$$14V = \left(11d^2 + 14s^2\right)h$$

The reference to an 'earlier' division by 14 presumably refers to the calculation of the volume of a cylinder, $V = \pi d^2 h / 4 \approx 11 d^2 h / 14$. This calculation occurs later in the text, translated in Section 3.10.5 below.

Therefore these are added together to make 25.

The 'diameters' of the square and the circle [s and d] are unequal in length, and when the 'diameters' are multiplied by themselves to make squares [s^2 and d^2], their sizes are not the same. For this reason the truncated [side of the] square, [$s-h =$] 2 *zhang* 9 *chi* 8 *cun* [$= 29.8$ *chi*], and the truncated diameter, [$d-h = (d-s) + (s-h) =$] 30 *zhang* 7 *cun* [$= 30.7$ *chi*], each form the side of a square. Thus [the sum of] the truncated [side of the] square, multiplied by itself [$(s-h)^2$], multiplied by 14, and the truncated diameter, multiplied by itself [$(d-h)^2$], multiplied by 11 and divided by 25, is [$K_1 =$] the Area for the Corner.

$$K_1 = \frac{14(s-h)^2 + 11(d-h)^2}{25} = 991.998 \ chi^2$$

and this is the *fangfa* [the coefficient of the linear term]. However, the two corner areas both have truncated numbers [$d-h$, $s-h$] as their sides, and this method is simpler.

◆ *3.10.4.2. Justification of the original calculation*

The extant text of the rest of the comment seems confused, and Qian Baocong's emendations correct not only obvious scribal errors but also the mathematical content. In an exception to our usual rule we do not always follow Qian Baocong's emendations.

In the following, consult Figure 17, p. 54.

Doubling the [side of the] small corner square [*MNOP*] [$s-h$] and adding the difference [$d-s$] gives [the length of] the gnomon[1]

倍小隅方加差為﹛短｜矩表[1]﹜以差乘之為﹛短｜矩﹜冪。一十一乘之，二十五而一。又﹛小隅方｜差﹜自乘之數即是方圓之隅同有﹛此此｜此﹜數，若二十五乘之，還須二十五除。直以﹛小隅方｜差﹜自乘加之，故不復乘除。又須倍二廉之差，一十一乘之，二十五而一，倍﹛二廉加之，故｜差加之﹜為廉法，不復二十五乘除之也。）

---

1 *Duan* 短, clearly a scribal error for *ju* 矩. Qian Baocong: *ju mao* 矩表, 'length of a gnomon'.

1 Li Huang 1832, *xia*, 39a.

[*EGIPON*], and multiplying by the difference [*d–s*] gives the area of the gnomon. This is multiplied by 11 and divided by 25.

Further, when [the side of] the small corner square[1] [*s–h*] is multiplied by itself, the corners of both the square and the circle have this same number. If it is multiplied by 25 it is necessary to divide by 25. The product of [the side of] the small corner square[2] [*s–h*] by itself is directly added; therefore one does not multiply and divide.

Compare (3.29).

It is further necessary to double the difference between the two sides [*d–s*], multiply by 11, and divide by 25. Adding twice the [difference between] the two sides[3] [*s–h*] gives the *lianfa*, without further multiplying and dividing by 25.

Compare (3.30).

3.10.5. Inverse procedure

還元術曰：倉方自乘，以高乘之，為實。圓徑自乘，以深乘之，一十一乘，一十四而一，為實。皆以斛法除之，即得容粟。

The method to return to the origin is: Multiply the [side of the] square of the granary by itself and multiply by the height to make a *shi* [dividend]. Multiply the diameter of the circle by itself, multiply by 11, and divide by 14 to make [another] *shi*. Divide each by the norm for *hu* to obtain the [amount of] grain that it holds.

3.10.6. Comment in smaller characters

斛法二尺五寸。

The norm for *hu* is 2 *chi* 5 *cun* [i.e. 2.5 *chi*^{3}].

---

1 *Xiao yu fang* 小隅方. Qian Baocong: *cha* 差, 'difference'.

2 *Xiao yu fang* 小隅方. Qian Baocong: *cha* 差, 'difference'.

3 The original has *er lian gu* 二廉故, 'two sides, therefore'; Qian Baocong replaces this with *cha* 差, 'difference'. We believe that the scribal error here is the omission of *cha* after *er lian*. We omit *gu*, 'therefore', in the translation.

3.11. Problem 11: Four square granaries and three circular storage pits

3.11.1. The problem

In the following, consult Section 2.3.2 above and Figure 18, p. 56.

Suppose there is 16,348 *dan* 8 *dou* [= 16,348.8 *hu*] of grain. It is desired to make [for this amount of grain] 4 square granaries and 3 circular storage pits. The side of the square [granary] is to be [*d–s* =] 1 *zhang* [= 10 *chi*] less than the diameter of the circular [storage pit] and greater than the height by [*s–h* =] 5 *chi*. The norm for *hu* is 2 *chi* 5 *cun* [i.e. 2.5 *chi*³/*hu*]. The proportions are diameter 7, circumference 22 [i.e. $\pi \approx {}^{22}/_7$]. What are [the side of] the square [granaries, *s*], the height [*h*], and the diameter [of the circular storage pits, *d*]?

假令有粟一萬六千三百四十八石八斗。欲作方倉四，圓窖三，令高、深等。方面少於圓徑一丈，多於高五尺。斜法二尺五寸。率徑七、周二十二。問方、高、徑各多少？

3.11.2. Answer

[Side of the] square, [*s* =] 1 *zhang* 8 *chi* [= 18 *chi*];
height and depth, [*h* =] 1 *zhang* 3 *chi* [= 13 *chi*];
diameter of the circular [storage pit], [*d* =] 2 *zhang* 8 *chi* [= 28 *chi*].

答曰：
　方一丈八尺，
　高深一丈三尺，
　圓徑二丈八尺。

3.11.3. Method

Multiply the norm for *hu* by 14, multiply by the amount of grain, and divide by 89 to make the *shi* [the constant term in the equation].

$$V = 16{,}348.8\ hu \times 2.5\ chi^3/hu = 40{,}872\ chi^3$$

$$shi = \frac{14V}{89} = 6{,}429\ {}^{27}/_{89}\ chi^3$$

Double the 'greater than' [*s–h*], add the 'less than' [*d–s*], multiply by the amount 'less than' [*d–s*], multiply by 33, and divide by 89. Add to this the 'greater than' [*s–h*] multiplied by itself to make the *fangfa* [the coefficient of the linear term].

術曰：以一十四乘斜法，以乘粟數，如八十九而一，為實。倍多加少，以乘少數，三十三乘之，八十九而一。多自乘加之，為方法。又倍少數，以三十三乘之，八十九而一，倍多加之，為廉法，從。開立方除之，即高、深。各加差即方徑。

$$fangfa = \frac{33}{89}\left[2(s-h)+(d-s)\right](d-s)+(s-h)^2$$

$$= 99\,^{14}/_{89}\; chi^2$$

Further double the amount 'less than' [d–s], multiply by 33, and divide by 89. Double the 'greater than' [s–h] and add to make the *lianfa* [the coefficient of the quadratic term].

$$lianfa = 2(d-s)\frac{33}{89}+2(s-h) = 17\,^{37}/_{89}\; chi$$

Extract the cube root; this is the height and depth [h].

$$h^3 + \left[2(d-s)\frac{33}{89}+2(s-h)\right]h^2$$

$$+ \left[\frac{33}{89}\left[2(s-h)+(d-s)\right](d-s)+(s-h)^2\right]h = \frac{14V}{89}$$

$$h^3 + 17\,^{37}/_{89}\,h^2 + 99\,^{14}/_{89}\,h = 6{,}429\,^{27}/_{89}\; chi^3$$

$$h = 13\; chi$$

Add each of the differences; this gives [the side of] the square [s] and the diameter [d].

$$s = (s–h) + h = 18\; chi$$

$$d = (d–s) + s = 28\; chi$$

3.11.4. Comment in smaller characters

一十四乘斜法，以乘粟，為徑自乘及方自乘數與前同。今方倉四即四因十四，圓窖三即三因十一，并之為八十九，而一。此{壅｜截}徑一丈五尺，{壅｜截}方五尺，以高為立方。自外意同前。

Multiplying the norm for *hu* by 14 and multiplying by the [amount of] grain gives the diameter [d] multiplied by itself and [the side of] the square, as in the previous [problem]. In this case there are four square granaries, so the 14 is multiplied by 4, and three circular storage pits, so the 11 is multiplied by 3; adding these gives 89, to be divided.

Here the truncated diameter is [d–h = (d–s)+(s–h) =] 1 *zhang* 5 *chi* [= 15 *chi*], the truncated [side of the] square is [s–h =] 5 *chi*, and the height [h] is taken as the [side of the] cube [i.e., the unknown quantity in the cubic equation]. From the outside the intention is the same as in the previous [problem].

3.12. Problem 12: Another square granary and circular storage pit

3.12.1. The problem

In the following, consult Section 2.3.3 above and Figure 19, p. 57.

Suppose there is grain amounting to 3,072 *dan* [= *hu*]. It is desired to make 1 square granary and 1 circular storage pit. The diameter and [the side of] the square are equal [*d=s*]; [the side of] the square is greater than the depth of the storage pit by [*s–g =*] 2 *chi* and less than the height of the granary by [*h–s =*] 3 *chi*. Both are completely filled by the grain. (The proportions of the circle and the norm for *hu* are as in the previous [problem].)[1] What are [the side of] the square, the diameter, the height, and the depth [*s, d, h, g*]?

$$s–g = 2 \; chi$$
$$h–s = 3 \; chi$$

假令有粟三千七十二石。欲作方倉一,圓窖一,令徑與方等,方多於窖深二尺,少於倉高三尺,盛各滿中而粟適盡。(圓率、斛法竝與前同。)問方、徑、高、深各多少?

3.12.2. Answer

[Side of] square and diameter, each [*s = d =*] 1 *zhang* 6 *chi* [= 16 *chi*];
height, [*h =*] 1 *zhang* 9 *chi* [= 19 *chi*];
depth, [*g =*] 1 *zhang* 4 *chi* [=14 *chi*].

答曰:
　方、徑各一丈六尺,
　高一丈九尺,
　深一丈四尺。

3.12.3. Method

Multiply the [amount of] grain by [14×2.5 =] 35 and divide by 25 to make [$K_1 =$] the Proportion.

$$V = 3{,}072 \; hu \times 2.5 \; chi^3/hu = 7{,}680 \; chi^3$$

$$K_1 = \frac{14V}{25} = 4{,}300.8 \; chi^3$$

術曰:三十五乘粟,二十五而一為率。多自乘,以并多少乘之,以乘一十四,如二十五而一,所得,以減率,餘,為實。并多少,以乘多,倍之,乘一十四,如二十五而一。多自乘加之,為方法。又并多少,以乘一十四,如二十五而一,倍多加之,為廉法,從。開立方除之,即窖深。各加差,即方、徑、高。

---

1   Comment in smaller characters. The values are $\pi \approx {}^{22}/_7$ and the norm for *hu* = 2.5 *chi³/hu*.

Multiply the 'greater than' [s–g] by itself; multiply by the sum of the 'greater than' [s–g] and the 'less than' [h–s], multiply by 14, and divide by 25. Subtract the result from the Proportion [K_1]; the difference is the *shi* [the constant term in the equation].

$$shi = K_1 - \frac{14(s-g)^2\left[(s-g)+(h-s)\right]}{25} = 4,289.6 \; chi^3$$

Add together the 'greater than' [s–g] and the 'less than' [h–s], multiply by the 'greater than' [s–g], and double it. Multiply by 14 and divide by 25. Multiply the 'greater than' [s–g] by itself and add to this to make the *fangfa* [the coefficient of the linear term].

$$fangfa = \frac{14 \times 2\left[(s-g)+(h-s)\right](s-g)}{25} + (s-g)^2$$

$$= 15.2 \; chi^2$$

Further add together the 'greater than' [s–g] and the 'less than' [h–s], multiply by 14, and divide by 25. Double the 'greater than' [s–g] and add it to make the *lianfa* [the coefficient of the quadratic term].

$$lianfa = \frac{14\left[(s-g)+(h-s)\right]}{25} + 2(s-g) = 6.8 \; chi$$

Extract the cube root; this is the depth of the storage pit [g].

$$g^3 + \left\{\frac{14\left[(s-g)+(h-s)\right]}{25} + 2(s-g)\right\} g^2$$

$$+ \left\{\frac{14 \times 2\left[(s-g)+(h-s)\right](s-g)}{25} + (s-g)^2\right\} g$$

$$= K_1 - \frac{14(s-g)^2\left[(s-g)+(h-s)\right]}{25}$$

$$g^3 + 6.8 \; g^2 + 15.2 \; g = 4,289.6 \; chi^3$$

$$g = 14 \; chi$$

Add each of the differences to obtain [the side of] the square [s], the diameter [d], and the height [of the granary, h].

$$s = d = g + (s-g) = 16 \; chi$$

$$h = g + (s-g) + (h-s) = 19 \; chi$$

3.12.4. *Comment in small characters*

This comment relates the method to that of the two previous problems.

The truncated height is [*h*–*g* =] 5 *chi*, the truncated diameter or [side of the] square is [*s*–*g* =] 2 *chi*, and the depth [*g*] is the [side of the] cube [i.e. the unknown to be solved for]. The norm for *hu* [2.5 *chi³/hu*] is multiplied by 14; therefore the [amount of] grain is multiplied by 35.

The 'greater than' [*s*–*g*] multiplied by itself, multiplied by the sum of the 'greater than' [*s*–*g*] and the 'less than' [*h*–*s*] is [K_2 =] the Corner Volume for the Truncated Height.

$$K_2 = (s-g)^2 \big[(s-g)+(h-s)\big] = 20 \ chi^3$$

(That is, the two, the *lian* and the *fang*, are each 2 *chi* [= *s*–*g*], and the length is 5 *chi* [= *h*–*g*].)[?][1] From the outside the intention is the same as in the previous [problem].

截高五尺，{塹|截}徑及方二尺，以深為立方。十四乘斜法，故三十五乘粟。多自乘幷多少乘之為截高隅積{減率餘|}，即二{方|}廉，{|方}各二尺，長五尺。自外意旨皆與前同。

---

1 Obscure sentence.

3.13. Problem 13: Two storage pits, one square and one circular

假令有粟五千一百四十五石。欲作方窖、圓窖各一，令口小底大，方面於圓徑等，兩深亦同，其深少於下方七尺，多於上方一丈四尺，盛各滿中而粟適盡。（圓率、斛法並與前同）問：方、徑、深各多少？

答曰：
　上方徑各七尺，
　下方徑各二丈八｛寸1尺｝，
　深各二丈一尺。

術曰：以四十二乘斛法，以乘粟，七十五而一，為方亭積。今方差自乘，三而一，為隅陽冪。以截1多乘之，減積，餘，為實。以多乘差，加冪為方法。多加差為廉法，從。開立方除之，即上方。加差即合所問。

---

1 In the exactly parallel passage in Problem 14 (see fn. 1 in the Chinese text in Section 3.14.3 below, p. 193) this character is excised by both Qian Baocong and Guo and Liu. That neither excises it here is probably due to an oversight; however, the fact that it occurs in both places suggests that it is not a simple scribal error, and should not be excised in either of the two passages. On the other hand, the inclusion or excision of the character makes no difference to the mathematics of the passages, so in our translation we simply follow Qian Baocong in both cases.

3.13.1. The problem

In the following, consult Section 2.1.3 above and Figure 13, p. 45.

Suppose there are 5,145 *dan* [= *hu*] of grain. It is desired to make a square storage pit and a circular storage pit, one of each. The mouth is to be smaller and the bottom larger, with the side of the square equal to the diameter of the circle, and the depths also equal. The depth [h] is to be less than [the side of] the lower square [a] by [$a-h =$] 7 *chi* and greater than [the side of] the upper square [b] by [$h-b =$] 1 *zhang* 4 *chi* [= 14 *chi*]. When filled they contain the grain exactly. (The proportion of the circumference and the norm for *hu* are the same as in the previous [problem].)[1] What are [the sides of] the squares [b, a], the diameters of the circles [b, a], and the depths [h]?

3.13.2. Answer

[The side of] the upper square, and the [upper] diameter, each [$b =$] 7 *chi*;
[the side of] the lower square, and the [lower] diameter, each [$a =$] 2 *zhang* 8 *chi* [= 28 *chi*];
the depths, each [$h =$] 2 *zhang* 1 *chi* [= 21 *chi*].

3.13.3. Method

Multiply the norm for *hu* by 42, multiply by [the amount of] grain, and divide by 75 to make [$V_{fangting} =$] the volume of the *fangting*.

$$V = 5,145 \ hu \times 2.5 \ chi^3/hu = 12,862.5 \ chi^3$$

$$V_{fangting} = \frac{42}{75}V = 7,203 \ chi^3$$

---

1 Comment in smaller characters. The values are $\pi \approx {}^{22}/_7$ and the norm for *hu* = 2.5 chi^3 / *hu*.

Multiply the difference between [the sides of] the squares [$a–b = (a–h) + (h–b)$] by itself and divide by three to make [$K_1 =$] the Area for the Corner *Yang*[*ma*].

$$K_1 = \frac{(a-b)^2}{3} = 147 \; chi^2$$

Multiply this by the truncated 'greater than' [$h–b$] and subtract from the volume [of the *fangting*, $V_{fangting}$]; the difference is the *shi* [the constant term of the equation].

$$shi = V_{fangting} - K_1 \, (h-b) = 5,145 \; chi^3$$

Multiply the 'greater than' [$h–b$] by the 'difference between' [$a–b$] and add the Area [for the Corner *Yangma*, K_1] to make the *fangfa* [the coefficient of the linear term].

$$fangfa = (a-b)(h-b) + K_1 = 441 \; chi^2$$

The sum of the 'greater than' [$h–b$] and the 'difference between' [$a–b$] is the *lianfa* [the coefficient of the quadratic term].

$$lianfa = (a-b)+(h-b) = 35 \; chi$$

Extract the cube root; this is [the side of] the upper square [b].

$$b^3 + \left[(a-b)+(h-b)\right]b^2$$
$$+ \left[(a-b)(h-b)+K_1\right]b = V_{fangting} - K_1 \, (h-b)$$

$$b^3 + 35b^2 + 441b = 5,145 \; chi^3 \qquad (3.31)$$

$$b = 7 \; chi$$

Add the differences; this gives the answer required.

$$h = b + (h-b) = 21 \; chi$$
$$a = h + (a-h) = 28 \; chi$$

3.13.4. Comment in smaller characters

In the case of a *fangting*, [the sides of] the upper and lower squares [b, a] are multiplied together; further, each is multiplied by itself; [the three results] are added and multiplied by the height [h] to

凡方亭，上、下方相乘，又{命｜各}自乘，幷以乘高，為虛。命三而一，為方亭積。若圓亭上、下徑相乘，又各自乘，幷以乘高，為虛。又十一乘之，四十二而一，為圓亭積。今方圓二積幷在一處，故以四十二復乘之，即得圓虛十一，方虛十四，凡二十五而一，得一虛之積。又三除虛積，為方亭實。乃依方{高｜亭}覆問法，見上、下方差及高差與積求上下方高術入之，故三乘，二十五而一。

make [$K_2 =$] the Space [xu 虛]. Dividing by 3 makes [$V_{fangting} =$] the volume.

$$K_2 = \left(ab + a^2 + b^2\right)h$$

$$V_{fangting} = \frac{K_2}{3}$$

In the case of a *yuanting*, [in the same way] the upper and lower diameters [b, a] are multiplied together; each is multiplied by itself; [the three results] are added and multiplied by the height [h] to make [$K_2 =$] the Space. Further multiplying by 11 and dividing by 42 makes [$V_{yuanting} =$] the volume of the *yuanting*.

$$V_{yuanting} = \frac{\pi}{4}\frac{K_2}{3} \approx \frac{11K_2}{42}$$

In the present case the volumes of a *fang*[*ting*] and a *yuan*[*ting*] are added together. Therefore, when [the total volume, V] is in return multiplied by 42, the result is 11 of the *yuan*[*ting*] Space and 14 of the *fang*[*ting*] Space, and when the whole is divided by 25, one Space volume [K_2] is obtained.

$$42V \approx 11K_2 + 14K_2$$

$$K_2 = \frac{3 \times 4V}{4 + \pi} \approx \frac{42V}{25} = 21{,}609 \ chi^3$$

Further dividing by 3 gives the volume [*shi* 實][1] of a *fangting*.

$$V_{fangting} = \frac{K_2}{3} = 7{,}203 \ chi^3$$

Therefore the method is sought from the *fangting*. The method is used by which, from the difference between [the sides of] the upper and lower squares [$a–b$], the height difference [$h–b$], and the volume [V], [the sides of] the upper and lower squares [a, b] and the height [h] are calculated [Problem 7, Section 3.7.3 above]. This is why there is a division by the product of 3 and 25.

---

1 The word *shi* often means 'volume' (and sometimes 'area') in other Chinese mathematical texts (see e.g. Chemla and Guo 2004: 977–979), but this seems to be the only use of the word in this meaning in *Jigu suanjing*, where it usually means the constant term in a polynomial equation and sometimes means 'dividend'.

3.14. Problem 14: Six square and four circular storage pits

3.14.1. The problem

In the following, consult Section 2.1.3.4 above and Figure 13, p. 45.

Suppose there are 26,342 *dan* 4 *dou* [= 26,342.4 *hu*] of grain. It is desired to make 6 square and 4 circular storage pits. [In each pit] the mouth is to be smaller and the bottom larger, with the side of the square equal to the diameter of the circle, and the depths also equal. The depth [*h*] is to be less than [the side of] the lower square [*a*] by [*a–h* =] 7 *chi* and greater than [the side of] the upper square [*b*] by [*h–b* =] 1 *zhang* 4 *chi* [= 14 *chi*]. When filled they are to contain the grain exactly. (The proportion of the circumference and the norm for *hu* are the same as in the previous [problem].)[1] What are [the sides of] the upper and lower squares [*a*, *b*] and the depth [*h*]?

3.14.2. Answer

[The side of] the upper square of the square storage pit, [*b* =] 7 *chi*;

[the side of] the lower square, [*a* =] 2 *zhang* 8 *chi* [= 28 *chi*];

depth, [*h* =] 2 *zhang* 1 *chi* [= 21 *chi*].

The upper and lower diameters of the circular storage pit [*a*, *h*] are the same as [the upper and lower sides of] the square storage pit.

3.14.3. Method

Multiply the norm for *hu* by 42, multiply by [the amount of] grain, and divide by 384 to make [$V_{fangting}$ =] the volume of [each] *fangting* in [cubic] *chi*.

$$V = 26{,}342.4 \ hu \times 2.5 \ chi^3/hu = 65{,}856 \ chi^3$$

---

1 Comment in smaller characters. The values are $\pi \approx {}^{22}/_7$ and the norm for *hu* = 2.5 *chi³/hu*.

假令有粟二萬六千三百四十二石四斗。欲作方窖六，圓窖四，令口小、底大，方面與圓徑等，其深亦同。今深少於下方七尺，多於上方一丈四尺。盛各滿中，而粟適盡。圓率、斜法竝與前同。問上、下方、深數各多少？

答曰：
　方窖上方七尺，
　下方二丈八尺，
　深二丈一尺，
　圓窖上下{方|徑}{深||深}與方窖同。

術曰：以四十二乘斜法，以乘粟，三百八十四而一，為方亭積尺。今方差自乘，三而一，為隅陽幂。以{截|}[1]多乘之，以減積，餘為實。以多乘差，加幂為方法。又以多加差為廉法，從。開立方除之，即上方。加差即合所問。

---

1 Note fn. 1 in the Chinese text in Section 3.13.3 above, p. 190.

$$V_{fangting} = \frac{42V}{384} = 7{,}203 \; chi^3$$

Multiply the difference between [the sides of] the squares [$a-b = (a-h) + (h-b)$] by itself and divide by three to make [$K_1 =$] the Area for the Corner *Yang*[*ma*].

$$K_1 = \frac{(a-b)^2}{3} = 147 \; chi^2$$

Multiply this by the 'greater than' [$h-b$] and subtract from the volume [$V_{fangting}$]; the difference is the *shi* [the constant term of the equation].

$$shi = V_{fangting} - K_1 \, (h-b) = 5{,}145 \; chi^3$$

Multiply the 'greater than' [$h-b$] by the 'difference between' [$a-b$] and add the Area [for the Corner *Yangma*, K_1] to make the *fangfa* [the coefficient of the linear term].

$$fangfa = (a-b)(h-b) + K_1 = 441 \; chi^2$$

Further add the 'greater than' [$h-b$] to the 'difference between' [$a-b$] to make the *lianfa* [the coefficient of the quadratic term].

$$lianfa = (a-b)+(h-b) = 35 \; chi$$

Extract the cube root; this is [the side of] the upper square [b].

$$b^3 + \left[(a-b)+(h-b)\right]b^2$$
$$+ \left[(a-b)(h-b)+K_1\right]b = V_{fangting} - K_1 \, (h-b)$$
$$b^3 + 35b^2 + 441b = 5{,}145 \; chi^3 \tag{3.32}$$
$$b = 7 \; chi$$

Note that (3.32) is exactly the same as (3.31), p. 191 above.

Add the differences; this gives the answer required.

$$a = h + (a-h) = 28 \; chi$$
$$h = b + (h-b) = 21 \; chi$$

3.14.4. Comment in smaller characters

This comment follows on from the comment translated in Section 3.13.4 above.

Here the multiplication by 42 [gives] 4 times 11 times the Space [K_2] for the circular [pit] and 6 times 14 times the Space [K_2] for the square [pit]; dividing by the sum, [$4 \times 11 + 6 \times 14 =$] 128, makes one Space volume.

$$42V \approx 4 \times 11 K_2 + 6 \times 14 K_2$$

$$K_2 = \frac{3V}{4 \times \dfrac{\pi}{4} + 6} \approx \frac{42V}{128}$$

As before, dividing this result by 3 makes the volume [*shi ji* 實積][1] of [one] *fangting*. So this is calculated using the method of calculating [the dimensions of] a *fangting* from differences. This is why there is a division by the product of 3 and 128.

今以四十二乘。圓虛十一者四，方虛十四者六，合一百二十八虛除之，為一虛之積。得者仍三而一為方亭實積。乃依方亭見差覆問求之，故三乘一百二十八除之。

---

1 Note fn. 1 in Section 3.13.4 above, p. 192.

3.15. Problem 15: A right triangle

In the following, consult Section 2.8 above and Figure 41, p. 102.

[In a right triangle] the area obtained by multiplying the base [a] by the leg [b] is [ab =] $706\,^{1}/_{50}$ and the hypotenuse [c] is greater than the base [a] by [c–a =] $36\,^{9}/_{10}$. How large are the three quantities?

Answer:

base, [a =] $14\,^{7}/_{20}$

leg, [b =] $49\,^{1}/_{5}$

hypotenuse, [c =] $51\,^{1}/_{4}$

Method: Multiply the area [ab] by itself and divide by twice the difference [2(c–a)] to make the *shi* [the constant term].

$$shi = \frac{(ab)^2}{2(c-a)} = 6{,}754\,^{129}/_{500}$$

Halve the difference [c–a] to make the *lianfa* [the quadratic coefficient].

$$lianfa = \frac{(c-a)}{2} = 18\,^{9}/_{20}$$

Extract the cube root; this is the base [a].

$$a^3 + \frac{(c-a)}{2}a^2 = \frac{(ab)^2}{2(c-a)}$$

$$a^3 + 18\,^{9}/_{20}\,a^2 = 6{,}754\,^{129}/_{500}$$

$$a = 14\,^{7}/_{20}$$

Add the hypotenuse difference [c–a]; this is the hypotenuse [c]. Divide the product [ab] by the base [a]; this is the leg [b].

$$b = \frac{ab}{a} = 49\,^{1}/_{5}$$

$$c = a + (c-a) = 51\,^{1}/_{4}$$

假令有句股相乘冪七百六、五十分之一，弦多於句三十六、十分之九。問三事各多少？

答曰：

句十四、二十分之七，

股四十九、五分之一，

弦五十一、四分之一。

術曰：冪自乘，倍多數而一，為實。半多｛｜數為｝廉法，從。開立方除之，即句。以弦多｛｜數加之｝，即弦。以句除冪，即股。

3.15.1. Comment in smaller characters

The product of the base and the leg, multiplied by itself [$(ab)^2$], is the 'volume' [*ji* 積] obtained by multiplying the area of [the square on] the base by the area of [the square on] the leg [a^2b^2].

$$(ab)^2 = a^2b^2$$

In the following, consult Figure 42, p. 103.

Therefore dividing by the doubled difference between the base and the hypotenuse [$2(c{-}a)$] gives the base and the halved difference lined up [$a + (c{-}a)/2$] multiplied by the area of [the square on] the base [a^2] to make a box [*fang* 方].

$$\frac{(ab)^2}{2(c-a)} = a^2\left(a + \frac{c-a}{2}\right)$$

This is why the difference is halved to make the *lianfa*, and the extraction of the cube root is carried out.

句股相乘冪自{　|乘，即}句冪乘股冪{　|之積。故}以倍句弦差而一，得一句與半差{　|相連，乘|再乘得[1]}句冪為方。故半差為廉{|法}，從，開立方除之。

_____

1 Zhang Dunren 1803.

3.16. Problem 16: A right triangle

假令有句股相乘冪四千三十六、五分之{ |一，股}少於弦六、五分之一。問弦多少？

答曰：弦一百一十四、十分之七。

術曰：冪自乘，倍少數而一，為實。半少為廉法，從。開立方除之，即股。加差，即弦。

In the following, consult Section 2.8 above and Figure 41, p. 102.

[In a right triangle], the area obtained by multiplying the base [a] by the leg [b] is [$ab=$] 4,036 $^1/_5$. The leg is less than the hypotenuse [c] by [$c-b =$] 6 $^1/_5$. How large is the hypotenuse?

Answer: The hypotenuse is [$c=$] 114 $^7/_{10}$.

Method: Multiply the area [ab] by itself and divide by twice the difference [$c-b$] to make the *shi* [the constant term].

$$shi = \frac{(ab)^2}{2(c-b)} = 1{,}313{,}783\,^1/_{10}$$

Halve the difference [$c-b$] to make the *lianfa* [the quadratic coefficient].

$$lianfa = \frac{c-b}{2} = 3\,^1/_{10}$$

Extract the cube root; this is the leg [b].

$$b^3 + \frac{c-b}{2}b^2 = \frac{(ab)^2}{2(c-b)}$$

$$b^3 + 3\,^1/_{10}\,b^2 = 1{,}313{,}783\,^1/_{10}$$

$$b = 108\,^1/_2$$

Add the difference [$c-b$] to obtain the hypotenuse [c].

$$c = b + (c-b) = 114\,^7/_{10}$$

3.17. Problem 17: A right triangle

In the following, consult Section 2.8 above and Figure 41, p. 102.

[In a right triangle] the area obtained by multiplying the base [a] by the hypotenuse [c] is [$ac =$] $1{,}337\,^1/_{20}$, and the hypotenuse is greater than the leg by [$c{-}b =$] $1\,^1/_{10}$. How large is the leg [b]?

Answer: [$b=$] $92\,^2/_5$.

Method: Multiply the area [ac] by itself and divide by twice the difference [$c{-}b$] to make [$V =$] a Volume.

$$V = \frac{(ac)^2}{2(c-b)} = 812{,}592\,^{11}/_{80}$$

Further multiply the difference [$c{-}b$] twice by itself, halve this, and subtract from the Volume [V]. The difference is the *shi* [the constant term].

$$shi = V - \frac{(c-b)^3}{2} = 812{,}591\,^{59}/_{125}$$

Further multiply the difference [$c{-}b$] by itself and double it to make the *fangfa* [the linear coefficient].

$$fangfa = 2(c-b)^2 = 2\,^{21}/_{50}$$

Further lay out the difference [$c{-}b$], multiply by 5, and divide by 2 to make the *lianfa* [the quadratic coefficient].

$$lianfa = \frac{5(c-b)}{2} = 2\,^3/_4$$

Extract the cube root. This is the leg [b].

$$b^3 + \frac{5(c-b)}{2}b^2 + 2(c-b)^2 b = V - \frac{(c-b)^3}{2}$$

$$b^3 + 2\,^3/_4\,b^2 + 2\,^{21}/_{50}\,b = 812{,}591\,^{59}/_{125}$$

$$b = 92\,^2/_5$$

假令有句弦相乘冪一千三百三十七、二十分之一，弦多於股一、十分之一。問股多少？

答曰：九十二、五分之二。

術曰：冪自乘，倍多而一，為立冪。又多再｛｜自｝乘，半之，減立冪，餘為實。又多數自乘，｛｜倍之，｝為方法。又置多數，五之，二而一，為廉｛｜法，從｝。開立方除之，即股。

句弦相乘冪自｛　｜乘，即句｝冪乘弦冪之
｛　｜積。故以倍｝股弦差而一，得一股與
｛半差　｜□□□□半差｝為方今多再自
乘半之為隅｛　｜□□□□□｝橫虛二立廉｛
　｜□□□□□□□□□｝倍之為從隅
｛　｜□□□□□□□□□□｝多為上廉
即二多｛　‖□□□□□□□□□[1]｝法故五
之二而一｛　｜□□□｝

1 Guo Shuchun and Liu Dun 1998, 2: 18.

3.17.1. Comment in smaller characters

Multiplying the area [ac] obtained by multiplying the base by the hypotenuse by itself gives the 'volume' [ji 積] obtained by multiplying the area of [the square on] the base [a^2] by the area of [the square on] the hypotenuse [c^2].

$$(ac)^2 = a^2c^2$$

From here on the text is very fragmentary.

Therefore dividing by twice the difference [c-b] gives one leg [b] and . . .

[5 characters missing] . . . to make a box [fang 方]. The difference [c−b], multiplied twice by itself and halved, is the corner-piece [yu 隅] . . .

[5 characters missing] . . . two standing side-pieces [li lian 立廉], transverse [heng 橫] and space[1] [xu 虛], . . .

[11 characters missing] . . . doubled to make the longitudinal corner-piece/s [zong yu 從隅], . . .

[11 characters missing] . . . the difference [c−b] makes the upper side-piece/s [shang lian 上廉] . . .

[9 characters missing] . . . fa 法 [coefficient?]. Therefore it is multiplied by 5 and divided by 2. . . .

[9 characters missing].

1 See fn. 1 in Section 2.1.3.3 above, p. 46.

3.18. Problem 18: A right triangle

In the following, consult Section 2.8 above and Figure 41, p. 102.

[In a right triangle] the area obtained by multiplying the leg [*b*] by the hypotenuse [*c*] is [*bc* =] . . .

[9 characters missing] . . . 3. The base [*a*] is less than the hypotenuse [*c*] by [*c–a* =] 50 . . .

[characters missing] . . . Answer: 6 . . .

[characters missing] . . . Method: Multiply the area [*bc*] by itself . . .

[11 characters missing] . . . multiply twice by itself and halve it . . .

[10 characters missing] . . . multiply and double it to make the *fangfa* [the linear coefficient]. . . .

[10 characters missing] . . . *lianfa* [the quadratic coefficient]. Extract the cube root . . .

[10 characters missing] . . . the area gives the leg.

假令有股弦相乘冪{ |四千七百三十九、五分之|四千四百二十八五分之[1]}三，句少於弦五十{ |四、五分之二。問股多少？|五問股多少[2]}

答曰：六{ |十八。|十六[3]}

術曰：冪自乘，{ |倍少數而一，為立冪。又少數}再自乘，半之，以{ |減立冪，餘為實。又少數自}乘，倍之，為方法。{ |又置少數，五之，二而一，為}廉法，從。開立方{ |除之，即句。加差即弦。弦除}冪，即股。

---

1 Nam Pyŏng-Gil.

2 Ibid.

3 Ibid.

3.19. Problem 19: A right triangle

假令有股弦相乘冪{ ｜七百二十六，句七、十分之｜五十分之三句一百分之¹}七。問股多少？

答曰：股二十{ ｜六、五分之二。｜五分之六²}

術曰：冪自{ ｜乘，為實。句自乘，為方法，從。開方}除之，所得，{ ｜又開方，即股}

---
1 Nam Pyŏng-Gil.
2 Ibid.

{｜□□□□□□□□□□□□□□}數亦是股{ ｜□□□□□□□□□□□}為長以股{ ｜□□□□□□□□□□}得股冪又開{ ｜。。。}股北分母常{‖須¹}{ ｜。。。}

---
1 Taibei MS (see Section 1.6 above).

In the following, consult Section 2.8 above and Figure 41, p. 102.

[In a right triangle] the area obtained by multiplying the leg [b] by the hypotenuse [c] is [$bc =$] . . .

[10 characters missing] . . . 7. How large is the leg [b]?

Answer: The leg is 20 . . .

[characters missing] . . . Method:

The area . . . by itself . . .

[12 characters missing] . . . divide it by . . . ; the result . . . [characters missing] . . .

3.19.1. Comment in smaller characters

[Comment:] [characters missing] . . . the number is also the leg [b] . . .

[characters missing] . . . to make the length [*chang* 長], . . . by the leg [b] . . .

[characters missing] . . . to obtain the area of [the square on] the leg [b]. Extract the . . . root . . .

[characters missing] . . . *gu bei fen mu chang* 股北分母常 [?]

3.20. Problem 20: A right triangle

In the following, consult Section 2.8 above and Figure 41, p. 102.

[In a right triangle] the leg [*b*] is 16 and [one] half . . .

[10 characters missing] . . . 14 [and] . . . twenty-fifths . . .

[characters missing] . . .

Answer: . . .

[characters missing] . . .

Method:

The area multiplied by itself . . .

[11 characters missing] . . . extract the . . . root and extract the square root of the result . . .

[characters missing]

假令有股十六、二分{ |之一，句弦相乘冪一百六}十四、二十五分{ |之十四。問句多少？}

答曰：{ |句八、五分之四。|八五分之四}

術曰：冪自乘{ |為實。股自乘，為方法，從。開方}除之，所得，又開方{ |即句。}

3.21. Colophon

秘書省
　緝古算經一卷一冊
　元豐七年九月　　日
　校定降授宣德郎秘書省校書郎臣葉
　　祖洽上進
　校定承議郎行秘書省校書郎臣王仲脩
　校定朝奉郎行秘書省校書郎臣錢長卿

　　　奉議郎守秘書丞臣韓宗古
　　　朝清郎試秘書少監臣孫覺
　　　降授朝散郎試秘書監臣趙彥若

Palace Library
Jigu suanjing, 1 *juan*, 1 volume
7th year of Yuanfeng 元豐, 9th month, — day [October 1084]

Collated and presented to the Throne by Your servant Ye Zuqia 葉祖洽, Collating Attendant in the Palace Library, honoured with the title Court Gentleman of Manifest Virtue.

Collation by Your servant Wang Zhongxiu 葉祖洽, acting Collating Attendant in the Palace Library, with the title Gentleman for Discussion.

Collation by Your servant Qian Changqing 錢長卿, acting Collating Attendant in the Palace Library, with the title Gentleman for Court Service.

Your servant Han Zonggu 韓宗古, probationary Aide in the Palace Library, with the title Court Gentleman Consultant.

Your servant Sun Jue 孫覺, acting Vice Director of the Palace Library, with the title Gentleman for Court Audiences.

Your Servant Zhao Yanjun 趙彥若, acting Director of the Palace Library, honoured with the title Gentleman for Closing Court.

> This colophon from the 1084 edition was copied without change into the 1213 edition and copied further into the two extant hand copies (see Section 1.5 above). The same colophon is seen in the two surviving books of that edition whose last pages survive, *Sunzi suanjing* and *Wucao suanjing*.[1]

---

1 *Song ke suanjing liuzhong* 1981, vols 3 and 4, last pages.

Appendices

Appendix 1: Cubic equations in Jigu suanjing

Section	Equation	Real roots
3.2.3.1	$x^3 + 170x^2 + 7{,}166\,\tfrac{2}{3}x = 1{,}677{,}666\,\tfrac{2}{3}$	70
3.2.3.2	$x^3 + 1{,}620x^2 + 850{,}500x = 146{,}802{,}375$	135
3.2.3.4	$x^3 + 276x^2 + 19{,}184x = 633{,}216$	24
3.2.3.5	$x^3 + 840x^2 = 4{,}459{,}000$	70, −76.42, −833.58
3.3.3.2	$x^3 + 5{,}004x^2 + 1{,}169{,}953\,\tfrac{1}{3}x$ $= 41{,}107{,}188\,\tfrac{1}{3}$	31, −278.80, −4,756.20
3.3.3.3	$x^3 + 3{,}298\,\tfrac{2}{31}x^2 + 2{,}474{,}941\,\tfrac{29}{31}x$ $= 23{,}987{,}761{,}548\,\tfrac{12}{31}$	1,920
3.4.3	$x^3 + 62x^2 + 696x = 38{,}448$	18
3.4.4.2	$x^3 + 594x^2 = 682{,}803$	33, −34.95, −592.05
3.4.4.3	$x^3 + 693x^2 + 42{,}471x = 683{,}665.488$	13.2, −83.12, −623.08
3.5.3.2	$x^3 + 1{,}728x^2 + 746{,}496x = 7{,}644{,}119{,}040$	1,440
3.6.3	$x^3 + 135x^2 + 4{,}550x = 285{,}000$	30
3.6.4	$x^3 + 315x^2 + 32{,}400x = 435{,}888$	12
3.7.3.1	$x^3 + 15x^2 + 66x = 360$	3
3.7.3.2.3	$x^3 + 18x^2 + 108x = 1{,}512$	6
3.8.3	$x^3 + 90x^2 = 839{,}808$	72
3.9.3	$x^3 + 30x^2 + 264x = 20{,}304$	18
3.9.4	$x^3 + 162x^2 + 8{,}748x = 215{,}784$	18
3.10.3	$x^3 + 60.392x^2 + 911.998x = 32{,}369.022$	15.5
3.11.3	$x^3 + 17\,\tfrac{37}{89}x^2 + 99\,\tfrac{14}{89}x = 6{,}429\,\tfrac{27}{89}$	13
3.12.3	$x^3 + 6.8x^2 + 15.2x = 4{,}289.6$	14
3.13.3	$x^3 + 35x^2 + 441x = 5{,}145$	7
3.14.3	$x^3 + 35x^2 + 441x = 5{,}145$	7
3.15	$x^3 + 18\,\tfrac{9}{20}x^2 = 6{,}754\,\tfrac{129}{500}$	$14\,\tfrac{7}{20}$
3.16	$x^3 + 3\,\tfrac{1}{10}x = 1{,}313{,}783\,\tfrac{1}{10}$	$108\,\tfrac{1}{2}$
3.17	$x^3 + 2\,\tfrac{3}{4}x^2 + 2\,\tfrac{21}{50}x = 812{,}591\,\tfrac{59}{125}$	$92\,\tfrac{2}{5}$

Appendix 2: Computer program to illustrate classical Chinese root extraction

This program extracts the positive root of a Wang Xiaotong cubic (see Section 1.5 and Box 2 above, p. 29) by the Chinese version of Horner's Method. It is written in the Basic programming language, and has been tested using the Chipmunk interpreter.[1] Basic is hardly to be recommended for serious programming, but it is simple enough that readers who have learned any programming language should be able to understand the program without more explanation.

```
10 ' Initialization.
20 '
30 epsilon = 0.001 ' Desired precision of result.
40 root = 0
50 '
60 ' Subroutine for reducing the roots of the equation by the value of a proposed digit.
70 '
80 sub chu(dv)
90     a = a+dv
100    b = b+a*dv
110    c = c-b*dv
120    a = a+dv
130    b = b+a*dv
140    a = a+dv
150    end sub
160 '
170 ' Ask for the coefficients of the equation.
180 '
190 print "coefficients"
200 input a,b,c
210 '
220 ' Propose a digit.
230 '
240 cq = c^(1/3)
250 order = 10^floor(log10(cq)+1)
260 d = 1
270 '
280 ' Reduce the roots of the equation by the value of the proposed digit.
290 '
300 chu(d*order)
310 '
320 ' If the proposed digit was too large, undo the previous action, propose a
321 ' smaller digit, and go back to try again.
```

1   www.nicholson.com/rhn/basic/.

```
330 '
340 if c < 0
350    chu(-d*order)
360    d = d-1
370    if d = 0
380      order = order/10
390      d = 9
400    endif
410    goto 300
420    endif
430 '
440 ' Add the value of the digit to the accumulated result.
450 '
460 root = root+d*order
470 '
480 'If the desired precision has not been reached, continue proposing digits.
490 '
500 if c > epsilon then goto 240
510 '
520 ' Print the final result and exit.
530 '
540 print "root=", root
550 end
```

Appendix 3: Computer program to illustrate classical Chinese root extraction with fractional coefficients

```
100 ' Initialization.
110 '
120 epsilon = 0.001
130 root = 0
140 '
150 'Subroutine for reducing the roots of the equation by the value of a
151 ' proposed digit.
160 '
170 sub chu(dv)
180    a = a+dv
190    b = b+a*dv
200    c = c-b*dv
210    a = a+dv
220    b = b+a*dv
230    a = a+dv
240    end sub
250 '
260 ' Ask for the coefficients of the equation.
270 '
280 print "coefficients: integer, numerator, denominator"
290 print "a:"
300 input a,anum,aden
310 print "b:"
320 input b,bnum,bden
330 print "c:"
340 input c,cnum,cden
350 '
360 ' Form improper fractions using a common multiple of the denominators.
370 '
380 if aden <> bden or aden <> cden then
390    anum = anum*bden*cden
400    bnum = bnum*aden*cden
410    cnum = cnum*aden*bden
420    divisor = aden*bden*cden
430 else
440    divisor = aden
450 endif
460 bnum = bnum*divisor
470 cnum = cnum*divisor^2
480 a = a*divisor+anum
```

```
490 b = b*divisor^2+bnum
500 c = c*divisor^3+cnum
510 '
520 ' Propose a digit.
530 '
540 cq = c^(1/3)
550 order = 10^floor(log10(cq)+1)
560 d = 1
570 '
580 ' Reduce the roots of the equation by the value of the proposed digit.
590 '
600 chu(d*order)
610 '
620 ' If the proposed digit was too large, undo the previous action, propose a
621 ' smaller digit, and go back to try again.
630 '
640 if c < 0
650    chu(-d*order)
660    d = d-1
670    if d = 0
680      order = order/10
690      d = 9
700    endif
710    goto 600
720    endif
730 '
740 ' Add the value of the digit to the accumulated result.
750 '
760 root = root+d*order
770 '
780 ' If the desired precision has not been reached, continue proposing digits.
790 '
800 if c > epsilon then goto 540
810 '
820 ' Print the final result and exit.
830 '
840 print "root=", root/divisor
850 end
```

Bibliography

All URLs cited in this book were verified in April 2017.

Balazs, Etienne. 1953. 'Études sur la société et l'économie de la Chine médiévale. I. Le traité économique du "Souei-Chou"'. *T'oung Pao* 42.3/4: 113–329. www.jstor.org/stable/4527352

——. 1964. 'History as a guide to bureaucratic practice'. In *Chinese civilization and bureaucracy: Variations on a theme*, edited by Arthur F. Wright. New Haven and London: Yale University Press, pp. 129–149. Translated by H. M. Wright.

Bao Tingbo 鮑廷博 ed. 1780. *Jigu suanjing* 缉古算經 (Continuation of ancient mathematics). (*Zhibuzuzhai congshu* 知不足齋叢書 edn).

Biot, Édouard. 1851. *Le Tcheou-li ou Rites des Tcheou*. 3 vols. Paris: Imprimerie Nationale. Facs. repr. Peking: Wen Tien Ko, 1940.

Bréard, Andrea. 1999. *Re-Kreation eines mathematischen Konzeptes im chinesischen Diskurs: 'Reihen' vom 1. bis 19. Jahrhundert*. (Boethius: Texte und Abhandlungen zur Geschichte der Mathematik und der Naturwissenschaften, 42). Stuttgart: Franz Steiner Verlag.

——. 2002. 'Problems of pursuit: Recreational mathematics or astronomy?' In *From China to Paris: 2000 years' transmission of mathematical ideas*, edited by Yvonne Dold-Samplonius *et al.* Stuttgart: Franz Steiner Verlag.

Chemla, Karine. 1982. 'Étude du livre "Reflets des mesures du cercle sur la mer"'. Unpublished dissertation, University of Paris XIII.

——. 1994. 'Similarities between Chinese and Arabic Mathematical Writings: (I) Root extraction'. *Arabic Sciences and Philosophy* 4: 207–266. journals.cambridge.org/abstract_S0957423900001235

——. 2010. 'Changes and continuities in the use of diagrams *tu* in Chinese mathematical writings (third century to fourteenth century) [I]'. *East Asian Science, Technology and Society* 4.2: 303–326. easts.dukejournals.org/citmgr?gca=ddeasts;4/2/303

Chemla, Karine, and Guo Shuchun. 2004. *Les neuf chapitres: Le classique mathématique de la Chine ancienne et ses commentaires*. Paris: Dunod.

Cullen, Christopher. 1993. 'Chiu chang suan shu'. In *Early Chinese texts: A bibliographical guide*, edited by Michael Loewe. Berkeley: Society for the Study of Early China and Institute of East Asian Studies, University of California, pp. 16–23.

——. 1996. *Astronomy and mathematics in ancient China: the Zhou bi suan jing*. (Needham Research Institute studies, 1). Cambridge: Cambridge University Press.

Dai Zhen 戴震, ed. 1777. *Jigu suanjing* 缉古算經 (*Weibo Xie* 微波榭 edn). www.scribd.com/doc/46794333

Du Shiran 杜石然. 1989. 'Suanchou tanyuan' 算籌探源 (The origin of calculating rods). *Zhongguo Lishi Bowuguan guankan* (Journal of the Museum of Chinese History) 中国历史博物馆馆刊 12: 28–36.

Guo Shirong 郭世荣. 1994. '"Jigu suanjing" zao yangguantai ti xin jie' 《缉古算经》造仰观台题新解 ('A new study on the problem of constructing an observatory in Wang Xiaotong's *Ji gu suan jing*'). *Ziran kexue shi yanjiu* 自然科学史研究 ('Studies in the history of natural sciences') 13.2: 106–113.

Guo Shuchun 郭书春. 1997. '"Hefang tongyi · Suanfa men" chutan' 《河防通议·算法门》初探 ('An elementary study on the *Suan fa men* of *He fang tong yi*'). *Ziran kexue shi yanjiu* 自然科学史研究 ('Studies in the history of natural sciences') 16.3: 223–232.

Guo Shuchun 郭书春 and Liu Dun 刘钝, eds. 1998. *Suanjing shi shu* 算经十书 (Ten mathematical classics). 2 vols. Shenyang: Jiaoyu Chubanshe. www.scribd.com/doc/53778139

Guo Shuchun 郭书春, Chen Zaixin 陈在新, and Guo Jinhai 郭金海. 2006. *Jade mirror of the four unknowns*. 2 vols. (Library of Chinese classics, Chinese–English / Da Zhonghua wenku, Han–Ying duizhao 大中华文库，汉英对照). Shenyang: Liaoning Education Press / Liaoning Jiaoyu Chubanshe. *Siyuan yujian* 四元玉鉴 (1303 CE), by Zhu Shijie 朱世杰. Full text + modern Chinese translation by Guo Shuchun and English translation by Chen Zaixin. Revised and supplemented by Guo Jinhai.

Guo Shuchun 郭书春, Joseph W. Dauben, and Xu Yibao 徐义保. 2013. *Nine chapters on the art of mathematics* 九章筹术. 3 vols. (Library of Chinese classics, Chinese–English / Da Zhonghua wenku, Han–Ying duizhao 大中华文库，汉英对照) Shenyang: Liaoning Education Press / Liaoning Jiaoyu Chubanshe. *Jiuzhang suanshu*, 'A critical edition and English translation based upon a new collation of the ancient text and modern Chinese translation by Guo Shuchun; English critical edition and translation, with notes by Joseph W. Dauben and Xu Yibao.'

Guo Tao 郭涛. 1994. 'Shuxue zai gudai shuili gongcheng zhong de yingyong – "Hefang tongyi · Suanfa" de zhushi yu fenxi' 数学在古代水利工程中的应用—《河防通议·算法》的注释与分析 (The application of mathematics in ancient hydraulic engineering – commentary and analysis of the *Suanfa* section of *Hefang tongyi*). *Nongye kaogu* 农业考古 ('Agricultural archaeology'), 1994.1: 271–778, 285.

Gupta, R. C. 2010. 'India's Contributions to Chinese Mathematics Through the Eighth Century C.E.'. In *Ancient Indian Leaps into Mathematics*, edited by B. S. Yadav and M. Mohan. Springer Science+Business Media, pp. 33–44. 'Reprinted from Gaṇita Bhāratī, vol. 11: 38–49, 1989, with a change of title'.

Han shu. 1962. 漢書 (Standard history of the Han Dynasty, by Ban Gu 班固, 32–92 CE, critical edn). Beijing: Zhonghua Shuju 中华书局.

He Shaogeng 何绍庚. 1989. ' "Jigu suanjing" gougu ti yiwen shi bu' 《缉古算经》勾股题佚文试补 (Attempted reconstruction of the fragmentary text on right triangles in *Jigu suanjing*). *Zhongguo Lishi Bowuguan guankan* 中国历史博物馆馆刊 (Journal of the Museum of Chinese History) 12: 37–43, 54.

Heath, Thomas. 1921. *A history of Greek mathematics*. 2 vols., Oxford: Clarendon Press.

Hefang tongyi. 1782. 河防通議 (Comprehensive discussion of Yellow River conservancy). (*Siku quanshu* 四庫全書 edn). Edited by Shakeshi 沙克什 (1278–1351, also called Shansi 贍思). www.scribd.com/collections/3809180/

Herbert, P. A. 1988. *Examine the honest, appraise the able: Contemporary assessments of civil service selection in early Tang China*. (Faculty of Asian Studies monographs: New series, 10). Canberra: Faculty of Asian Studies, Australian National University.

Horner, W. G. 1819. 'A New Method of Solving Numerical Equations of All Orders, by Continuous Approximation'. *Philosophical Transactions of the Royal Society of London* 109: 308–335. www.jstor.org/stable/107508

Horng, Wann-Sheng 洪萬生. 2002. 'Sino-Korean transmission of mathematical texts in the 19th century: A case study of Nam Pyŏng-gil's *Kugo Sulyo Tohae*'. *Historia Scientiarum* 12: 87–99. Not seen. cat.inist.fr/?aModele=afficheN&cpsidt=14716065

Høyrup, Jens. 2013. *Algebra in cuneiform: Introduction to an Old Babylonian geometrical technique*. Berlin: Max-Planck-Institut für Wissenschaftsgeschichte.

Hu Daojing 胡道静, ed. 1962. Mengxi bitan jiaozheng 夢溪筆談校證 (Critical ed. of 'Dream Brook essays'). Shanghai. Facs. repr. Shanghai: Shanghai Guji Chubanshe, 1987.

——, Jin Liangnian 金良年, Hu Xiaojing 胡小静, Wang Hong 王宏, and Zhao Zheng 赵峥. 2008. *Brush talks from Dream Brook* 梦溪笔谈. 2 vols. (Library of Chinese classics Chinese–English). Chengdu: Sichuan Publishing Group and Sichuan People's Publishing House. 'Translated into modern Chinese by Hu Daojing, Jin Liangnian and Hu Xiaojing. Translated into English by Wang Hong and Zhao Zheng.'

Huang Juncai 黃俊才 (Chun-Tsai Huang). 2009. *Li Huang «Jigu suanjing kaozhu» zhi neirong fenxi* 李潢《緝古算經考注》之內容分析 (Analysis of the content of Li Huang's edition of *Jigu suanjing*). M.A. thesis, National Taiwan Normal University. handle.ncl.edu.tw/11296/ndltd/34091848210952812718.

Hucker, Charles O. 1985. *A dictionary of official titles in Imperial China*. Stanford: Stanford University Press.

Hulsewé, A. F. P. 1985. *Remnants of Ch'in law: An annotated translation of the Ch'in legal and administrative rules of the 3rd century B.C. discovered in Yün-meng Prefecture, Hu-pei Province, in 1975*. (Sinica Leidensia, 17). Leiden: Brill.

Jing Zhuyou 景竹友. 1993. 'Santai Xindexiang Dong Han yamu qingli jianbao' 三台新德乡东汉崖墓清理简报 (Clearing of Eastern Han cliff tombs at Xindexiang in Santai County, Sichuan). *Sichuan wenwu* 四川文物 (Sichuan cultural relics) 1993.5: 68–69 + inside front cover.

Jiu Tang shu. 1975. 舊唐書 (Older standard history of the Tang Dynasty, completed in 945 CE, critical edn). Shanghai: Zhonghua Shuju 中華書局.

Johnson, Wallace. 1979–1997. *The T'ang code*. 2 vols. (Studies in East Asian Law, Harvard University, 10). Princeton: Princeton University Press. 1. General principles. 2. Specific articles.

Kaiyuan zhanjing. 開元占經 (Astrological treatise of the Kaiyuan era). (*Siku quanshu* 四庫全書 edn).

Lam Lay-Yong [Lan Lirong 蓝丽蓉]. 1970. 'The geometrical basis of the ancient Chinese square-root method'. *Isis* 61.1: 92–102. www.jstor.org/stable/229151

——. 1977. *A critical study of the Yang Hui Suan Fa, a 13th-century Chinese mathematical treatise*. Singapore: Singapore University Press.

——. 1986. 'The development of polynomial equations in traditional China'. *Mathematical medley* (Singapore Mathematical Society) 14.1: 9–34.

——. 1987. 'Linkages: Exploring the similarities between the Chinese rod numeral system and our numeral system'. *Archive for history of exact sciences* 37: 365–392.

——. 1994. 'Jiu zhang suanshu 九章算术 (Nine chapters on the mathematical art): An overview'. *Archive for history of exact sciences* 47: 1–51.

Lam Lay-Yong [Lan Lirong 蓝丽蓉], and Ang Tian Se [Hong Tianci 洪天赐]. 2004. *Fleeting footsteps: Tracing the conception of arithmetic and algebra in ancient China*. Singapore: World Scientific Publishing. Revised edn.

Lam Lay-Yong [Lan Lirong 蓝丽蓉], and Shen Kangshen 沈康身. 1985. 'The Chinese concept of Cavalieri's principle and its applications'. *Historia Mathematica* 12.3: 219–228.

Li Di 李迪. 1999. 'Wang Xiaotong "Jigu suanjing"' 王孝通《缉古算经》 (Wang Xiaotong and his *Jigu suanjing*). In *Zhongguo shuxue shi daxi*: *4. Xijin zhi Wudai* 中国数学史大系 第四卷 西晋至五代, edited by Wu Wenjun 吴文俊 and Shen Kangshen 沈康身. Beijing: Beijing Shifan Daxue Chubanshe, pp. 196–218.

Li Huang 李潢, ed. [1832]. *Jigu suanjing kaozhu* 缉古算經考注 (Jing Yushu 靖玉樹, ed., *Lidai suanxue jicheng* 歷代算學集成, 1994). Colophon dated 1832. www.scribd.com/doc/43099474

Li Jin 栗劲. 1985. *Qin lü tonglun* 秦律通论 (A survey of Qin law). Ji'nan: Shandong Renmin Chubanshe.

Li Shengwu 李胜伍, and Guo Shuchun 郭书春. 1982. 'Shijiazhuang Dong Han mu ji qi chutu de suanchou' 石家庄东汉墓及其出土的算筹 (An Eastern Han tomb in Shijiazhuang, Hebei, and the calculating rods found in it). *Kaogu* 考古 ('Archaeology'), 1982.3: 255–256.

Li Yan 李严. 1954. 'Tang Song Yuan Ming shuxue jiaoyu zhidu' 唐宋元明數學教育制度 (The administration of mathematical education in the Tang, Song, Yuan, and Ming periods). In Zhong suan shi luncong 中算史論叢. Beijing: Kexue Chubanshe, vol. 4, pp. 238–280.

——. 1963a. *Zhongguo gudai shuxue shiliao* 中国古代数学史料 (Historical materials on ancient Chinese mathematics). 2nd edn Shanghai: Shanghai Kexue Jishu Chubanshe. 1st edn 1954.

——. 1963b. 'Gu jiujiubiao' 古九九表 (The ancient multiplication table). In *Zhongguo gudai shuxue shiliao* 中国古代数学史料, ed. by Li Yan. Shanghai: Shanghai Kexue Jishu Chubanshe, pp. 14–18.

——. 1963c. 'Chousuan zhidu' 籌算制度 (The system of calculating rods). In *Zhongguo gudai shuxue shiliao* 中国古代数学史料, ed. by Li Yan. Shanghai: Shanghai Kexue Jishu Chubanshe, pp. 14–18.

——. 1963d. 'Gu suan jieshi' 古算解释 (Explications of some ancient mathematics). In *Zhongguo gudai shuxue shiliao* 中国古代数学史料, ed. by Li Yan. Shanghai: Shanghai Kexue Jishu Chubanshe, pp. 76–80.

——. 1998. '«Jigu suanjing» jie' 《缉古算经》解 (Explications of items in *Jigu suanjing*). In *Li Yan Qian Baocong kexueshi quanji* 李俨钱宝琮科学史全集 (Collected studies in the history of mathematics by Li Yan and Qian Baocong). 10 vols., Shenyang: Liaoning Jiaoyu Chubanshe, vol. 3, pp. 128–138.

Li Yan and Du Shiran. 1987. *Chinese mathematics: a concise history*. Oxford: Clarendon. Translated by John N. Crossley and Anthony W.-C. Lun.

Libbrecht, Ulrich. 1973. *Chinese mathematics in the thirteenth century: The Shu-shu chiu-chang of Ch'in Chiu-shao*. (MIT East Asian science series, 1). Facs. repr. Mineola, NY: Dover, 2005.

——. 1982. 'Mathematical manuscripts from the Tunhuang caves'. In *Explorations in the history of science and technology in China*, edited by Li Guohao et al., Shanghai: Shanghai Chinese Classics Publishing House, pp. 203–229.

Lim, Tina Su Lyn, and Donald B. Wagner. 2013a. 'The Grand Astrologer's platform and ramp: Four problems in solid geometry from Wang Xiaotong's 'Continuation of ancient mathematics' (7th century AD)'. *Historia Mathematica* 40.1: 3–35. www.sciencedirect.com/science/article/pii/S0315086012000596

——. 2013b. 'Wang Xiaotong on right triangles: Six problems from 'Continuation of ancient mathematics' (7th century AD)'. *East Asian science, technology and medicine* 37: 12–35. Published 2016. www.eastm.org/index.php/journal/article/view/648/562

Lin Yanquan 林炎全. 2001. *Jigu suanjing tantao* 輯古算經探討 (A study of *Jigu suanjing*). Taiwan Sheng Zhongdeng Jiaoshi Yanxi Hui 台灣省中等教師研習會. www.scribd.com/doc/78851923

Liu Junwen 劉俊文, ed. 1983. *Tang lü shuyi* 唐律疏議 (Critical edition of 'Annotated Tang regulations'). Beijing: Xinhua Shuju.

Loewe, Michael. 2000. *A biographical dictionary of the Qin, Former Han and Xin periods (221 BC – AD 24)*. (Handbuch der Orientalistik / Handbook of Oriental Studies, IV: China, 60). Leiden / Boston / Köln: Brill.

Lu Liancheng 卢连成, Shi Xiezhong 时协中, and Mei Rong-zhao 梅荣照. 1976. 'Qianyang xian Xi Han mu zhong chutu suanchou' 千阳县西汉墓中出土算筹 (Calculating rods from a Western Han tomb in Qianyang County, Shaanxi). *Kaogu* 考古 ('Archaeology'), 1976.2: 85–88, 108 + Plate 1.

Luo Tengfeng 駱騰鳳 (1770–1841). 1993. 'Yi you lu' 藝游錄. In *Zhongguo kexue jishu dianji tonghui: Shuxue juan 5* 中国科学技术典籍通汇：数学卷 5 (Collected ancient texts on the history of science and technology in China: Mathematics, vol. 5). Zhengzhou: Henan Jiaoyu Chubanshe 河南教育出版社, pp. 141–211.

Martzloff, Jean-Claude. 1997. *A history of Chinese mathematics*. Tr. by Stephen S. Wilson. Berlin / Heidelberg: Springer-Verlag. 'Corrected second printing', 2006. Orig. *Histoire des mathémathiques chinoises*, Paris: Masson, 1987.

McMullen, David. 1988. *State and scholars in T'ang China*. (Cambridge studies in Chinese history, literature and institutions). Cambridge: Cambridge University Press.

Mei Rongzhao 梅荣照. 1966. 'Li Ye ji qi shuxue zhuzuo' 李冶及其数学著作 (Li Ye and his mathematical works). In *Song Yuan shuxue shi lunwenji* 宋元数学史论文集 (Essays on the history of mathematics in the Song and Yuan periods, 960–1368 CE), ed. by Qian Baocong 錢寶琮. Beijing: Kexue Chubanshe, pp. 104–148.

Mikami, Yoshio. 1913. *The development of mathematics in China and Japan*. (Abhandlungen zur Geschichte der mathematischen Wissenschaften, 30). Leipzig: Teubner.

Nam Pyŏng-Gil 南秉吉 (1820–1869) 1985. 緝古算段. (韓國科學技術史資料大系). Facs. repr. 1985. www.scribd.com/doc/27920990

Nan Qi shu. 1972. 南齊書 (Standard history of the Southern Qi Dynasty). Beijing: Zhonghua Shuju.

Needham, Joseph, and Wang Ling 王鈴. 1959. *Science and civilisation in China*. Volume 3: *Mathematics and the sciences of the heavens and the earth*. Cambridge: Cambridge University Press.

——, Wang Ling 王鈴, and Lu Gwei-djen. 1971. *Science and civilisation in China*. Vol. 4, part 3: *Civil engineering and nautics*. Cambridge: Cambridge University Press.

Nesselmann, G. H. F. 1842. *Versuch einer kritischen Geschichte der Algebra. Erster Theil: Die Algebra der Griechen*. Berlin: Reimer.

Pratt, Vaughan. 2014. 'Algebra'. In *The Stanford encyclopedia of philosophy* (spring 2014 edn), edited by Edward N. Zalta. plato.stanford.edu/archives/spr2014/entries/algebra/

Pu Qilong 浦起龍. 1961. *Shi tong tong shi* 史通通釋 (Critical edn of 'Generalities on history', by Liu Zhiji 劉知幾, 661–721). Hong Kong: Taiping Shuju. Facs. repr. 1977 of a typeset edn 1961 of an 18th cent. edn.

Qian Baocong 钱宝琮. 1958. *Zhongguo shuxue shihua* 中国数学史话 (Short history of mathematics in China). Beijing: Zhongguo Qingnian Chubanshe.

——, ed. 1963. *Suanjing shi shu* 算經十書 (Ten mathematical classics). Beijing: Zhonghua Shuju. www.scribd.com/doc/53797787

——. 1966a. 'Wang Xiaotong «Jigu suanjing» di er ti, di san ti shuwen shuzheng' 王孝通《缉古算经》第二题,第三题术文疏証 (Explication of problems 2 and 3 in Wang Xiaotong's *Jigu suanjing*). *Kexueshi jikan* 科学史集刊 (Studies in the history of science) 9: 31–52.

——. 1966b. *Song Yuan shuxue shi lunwenji* 宋元数学史论文集 (Studies of the history of mathematics in the Song and Yuan periods). Beijing: Kexue Chubanshe.

Rees, Paul K., and Fred W. Sparks. 1967. *College algebra*. 5th ed. New York: McGraw-Hill.

Rotours, Robert des. 1932. *Le traité des examens, traduit de la Nouvelle histoire des T'ang (chap. XLIV, XLV)*. (Bibliothèque de l'Institut des Hautes Études Chinoises, 2). Paris: Leroux.

——. 1947–48. *Traité des fonctionnaires et Traité de l'armée: traduits de la Nouvelle histoire des T'ang*. 2 vols. (Bibliothèque de l'Institut des Hautes Études Chinoises, 7). Leyde: Brill.

——. 1975. 'Le *T'ang liu tien* décrit-il exactement les institutions en usage sous la dynastie des T'ang?' *Journal asiatique* 263: 183–201.

Shen Kangshen 沈康身. 1964. 'Wang Xiaotong kaihe zhudi ti fenxi' 王孝通开河筑堤分析 (An analysis of Wang Xiaotong's problem of digging a canal and building a dyke). *Hangzhou daxue xuebao (Ziran kexue)* 杭州大学学报(自然科学) (Journal of Hang Zhou University (Natural sciences)) 2.4: 43–58.

Shen Kangshen 沈康身, John N. Crossley, and Anthony W.-C. Lun. 1999. *The nine chapters on the mathematical art: Companion and commentary*. Oxford and Beijing: Oxford University Press and Science Press.

Shisan jing zhushu. 1980. 十三經注疏 (The Thirteen Classics, with commentaries and further commentaries). 2

vols. Beijing: Zhonghua Shuju. Repr. of edn of Ruan Yuan 阮元, 1816.

Shu shu jiu zhang. 1842. 數書九章 (Mathematical treatise in nine sections). (*Yijiatang congshu* 宜稼堂叢書 edn). ctext.org/library.pl?if=en&&res=78705

——. 1936. 數書九章 (Mathematical treatise in nine sections). (*Congshu jicheng* 叢書集成 edn). www.guoxuedashi.com/guji/249128k/

Shuihudi. 1978. *Shuihudi Qin mu zhujian* 睡虎地秦墓竹简 (Texts written on bamboo found in a Qin-period tomb at Shuihudi in Yunmeng County, Hubei). Beijing: Wenwu Chubanshe.

Siu Man-Keung, and Alexeï Volkov. 1999. 'Official curriculum in traditional Chinese mathematics: How did candidates pass the examinations?' *Historia scientiarum* 9.1: 85–99.

Song ke suanjing liuzhong. 1981. 宋刻算经六种 (Six mathematical texts printed in the Song period). 6 vols. Beijing: Wenwu Chubanshe. Facsimile reprint of remains of the 'Ten mathematical canons' printed in 1213. Six thread-bound volumes in a case.

Sui shu. 1973. 隋書 (Standard history of the Sui Dynasty, completed 636 CE). 6 vols. Beijing: Zhonghua Shuju.

Taga Akigorō 多賀秋五郎. 1953. *Tōdai kyōiku shi no kenkyū* 唐代教育史の研究 (Studies on the history of education in the Tang period). Tōkyō: Fumaidō.

——. 1985. *The history of education in T'ang China.* Tr. by P. A. Herbert. Osaka: Osaka University. Summary translation of Taga 1953.

Taibai yinjing. 1854. 太白陰經 (Dark canon of the martial planet [by Li Quan 李筌 (Tang)]). (*Changsi Shushi congshu* 長思書室叢書 edn). play.google.com/store/books/details?id=G7QqAAAAYAAJ.

Tang liu dian. 唐六典 (Compendium on the administration of the Tang Dynasty). (*Siku quanshu* 四庫全書 edn). ctext.org/library.pl?res=6869, zh.wikisource.org/zh/唐六典

Tiansheng ling. 2006. *Tianyige cang Ming chaoben Tiansheng ling jiaozheng: Fu Tang ling fuyuan yanjiu* 天一閣藏明鈔本天聖令校證：附唐令復原研究 (The Ming hand-copy of 'Administrative regulations of the Tiansheng reign' preserved in the Tianyi Pavilion: With a study of the Tang regulations). 2 vols. Beijing: Zhonghua Shuju.

Tong dian. 1988. 通典 (Comprehensive documents). 5 vols. Beijing: Zhonghua Shuju.

Twitchett, Denis. 1979. 'The problem of sources'. In *The Cambridge history of China.* Vol. 3: *Sui and T'ang China, 589–906,* Part I, edited by Denis Twitchett. Cambridge: Cambridge University Press, pp. 38–47.

——. 1992. *The Writing of official history under the T'ang.* Cambridge: Cambridge University Press.

van der Waerden, B. L. 1983. *Geometry and algebra in ancient civilizations.* Berlin / Heidelberg / New York / Tokyo: Springer-Verlag.

Vogel, Kurt. 1968. *Neun Bücher arithmetischer Technik: Ein chinesisches Rechenbuch für den praktischen Gebrauch aus der frühen Hanzeit (220 v.Chr. bis 9 n.Chr.), übersetzt und erläutert.* (Ostwalds Klassiker der eksakten Wissenschaften, N. F., 4). Braunschweig: Vieweg.

Volkov, Alexeï. 1994. 'Calculation of π in ancient China: From Liu Hui to Zu Chongzhi'. *Historia Scientiarum* 4.2: 139–157.

——. 2012. 'Argumentation for state examinations: Demonstration in traditional Chinese and Vietnamese mathematics'. In *The history of mathematical proof in ancient traditions,* edited by Karine Chemla. Cambridge: Cambridge University Press, pp. 509–551.

Wagner, Donald B. 1978a. 'Doubts concerning the attribution of Liu Hui's commentary on the Chiu-chang suanshu'. *Acta Orientalia* 39: 199–212. donwagner.dk/LiuHui/LiuHui.html

——. 1978b. 'Liu Hui and Tsu Keng-chih on the volume of a sphere'. *Chinese science* 3: 59–79. donwagner.dk/SPHERE/SPHERE.html

——. 1979. 'An ancient Chinese derivation of the volume of a pyramid: Liu Hui, third century A.D.' *Historia mathematica* 6: 164–188. donwagner.dk/Pyramid/Pyramid.html

——. 2012a. 'Shen Gua and an ignorant editor on the length of an arc'. donwagner.dk/Shen-Gua-arc.htm

——. 2012b. 'Another ignorant editor? Note on a mathematical problem in a 14th-century Chinese treatise on river conservancy'. donwagner.dk/another.htm

——. 2013. 'Mathematics, astronomy, and the planning of public works in China, Han to Yuan'. donwagner.dk/SAW/SAW.html

——. 2017. 'The classical Chinese version of Horner's method: Technical considerations'. donwagner.dk/horner/horner.html.

Wang Chao 王超. 1984. 'Woguo gudai de xingzheng fadian – "Da Tang liu dian"' 我国古代的行政法典–《大唐六典》 (*Da Tang liu dian* – An ancient Chinese administrative law code). *Zhongguo shehui kexue* 中国社会科学 ('Social sciences in China') 1984.1: 115–142.

——. 1986. 'The *Six Codes of the Tang dynasty*: China's earliest administrative code'. *Social sciences in China* 1986.2: 113–150. Translation of Wang Chao 1984.

Wang Ling 王鈴 and Joseph Needham. 1955. 'Horner's method in Chinese mathematics: Its origins in the root-extraction procedures of the Han dynasty'. *T'oung Pao* 43.5: 345–401. www.jstor.org/stable/4527405

Wang Maohua 王茂华, Yao Jian'gen 姚建根, and Lü Wenjing 吕文静. 2012. 'Zhongguo gudai cheng chi gongcheng jiliang yu jijia chutan' 中国古代城池工程计量与计价初探 ('Measurement and Evaluation of City Wall and Moat Construction in Ancient China'). *Zhongguo keji shi zazhi* 中国科技史杂志 ('The Chinese journal for the history of science and technology') 33.2: 204–221.

Wang Qingjian 王青建. 1993. 'Shilun chutu suanchou' 试论出土算筹 ('An elementary exploration of unearthed counting-rods'). *Zhongguo keji shiliao* 中国科技史料 ('China historical materials of science and technology') 14.3: 3–11.

Wang Rongbin 王荣彬. 1990. 'Wang Xiaotong "Jigu suanjing" zizhu yiwen jiaobu' 王孝通《缉古算经》自注佚文校补 (Reconstruction of the missing text in Wang Xiaotong's commentary to his *Jigu suanjing*). In *Shuxue shi yanjiu wenji, di-yi-ji* 数学史研究文集，第一辑 (Studies in the history of mathematics, 1), edited by Li Di 李迪. Hohhot / Taibei: Neimenggu Daxue Chubanshe / Jiuzhang Chubanshe, pp. 50–55.

Wang Xiaoqin 汪晓勤. 1999. '"Zhui shu" zhong de "chumeng, fengting zhi wen" chutan' 《缀术》中的"刍甍、方亭之问"初探 ('Reflections on "the *chumeng* and *fangting* problems" in *Zhui shu*'). *Ziran kexue shi yanjiu* 自然科学史研究 ('Studies in the history of natural sciences') 18.1: 20–27.

Weisstein, Eric W. [1996?]. 'Cubic formula '. *MathWorld – A Wolfram Web Resource*. mathworld.wolfram.com/CubicFormula.html.

Xiao Qi 肖琦. 1988. 'Longxian Xi Han mu chutu suanchou' 陇县西汉墓出土算筹 (Calculating rods unearthed from a Western Han tomb in Longxian County, Shaanxi). *Kaogu yu wenwu* 考古与文物 ('Archaeology and cultural relics'), 1988.3: 108–109.

Xin Tang shu. 1975. 新唐書 (New standard history of the Tang dynasty, by Ouyang Xiu 欧阳修 [1007–1072]

and Song Qi 宋祁 [998–1061], critical edn). Shanghai: Zhonghua Shuju.

Yabuuti Kiyosi [Yabuuchi Kiyoshi 藪內清]. 1979. 'Researches on the *Chiu-chih li*: Indian astronomy under the T'ang dynasty'. *Acta Asiatica* 36: 7–47.

Yang Bojun 楊伯峻, ed. 1960. *Meng zi yizhu* 孟子譯注 (Critical edn and transl. of *Mencius*). Beijing: Zhonghua Shuju.

Ying, Jia-ming 英家銘. 2011. 'The *Kujang sulhae* 九章術解: Nam Pyŏng-Gil's reinterpretation of the mathematical methods of the *Jiuzhang suanshu*'. *Historia Mathematica* 38.1: 1–27. dx.doi.org/10.1016/j.hm.2010.04.001

Zhang Dunren 張敦仁, ed. 1803. *Jigu suanjing* 緝古算經 (*Congshu jicheng* 叢書集成 edn). www.scribd.com/doc/48275688/

Zhang Fukai (Fu-kai Chang) 張復凱. 2005. *Cong Nan Bingji (1820–1869) «Yigu yanduan» kan dong suan shi shang tianyuanshu yu jie gen fang zhi "duihua"* 從南秉吉 (1820~1869)《緝古演段》看東算史上天元術與借根方之「對話」 ('On the Dialogue between the Chinese *Tianyuan shu* and the Western *Jiegen fang*: A Korean Version by Nam Pyŏng-Gil (1820-1869)'). M.A. Thesis, National Taiwan Normal University. handle.ncl.edu.tw/11296/ndltd/39549827772884510360.

Zhang Pei 张沛. 1988. 'Suanchou de chansheng, fazhan ji qi xiang suanpan de yanbian' 算筹的产生、发展及其向算盘的演变 (The origin and development of calculation rods, and the development of the abacus). *Dongnan wenhua* 东南文化 ('Southeastern culture'), 1988.6: 130–138.

——. 1992. 'Suanchou shuoyuan' 算筹溯源 (The origin of counting rods). *Wenbo* 文博 ('Relics and museology'), 1992.3: 65–72.

——. 1996. 'Chutu suanchou kaolüe' 出土算筹考略 (Calculating rods from excavations). *Wenbo* 文博 ('Relics and museology'), 1996.4: 53–59.

Zhu Shida 朱世达 (tr.). 2007. *Taibai yinjing: Han Ying duizhao* 太白阴经：汉英对照. (Da Zhonghua wenku 大中华文库). Beijing: Junshi kexue chubanshe. 'Dark canon of the martial planet', by Li Quan 李筌 (Tang). Full text of *juan* 1–3, with modern Chinese and English translations.

Index and glossary

NIAS Press is the autonomous publishing arm of NIAS – Nordic Institute of Asian Studies, a research institute located at the University of Copenhagen. NIAS is partially funded by the governments of Denmark, Finland, Iceland, Norway and Sweden via the Nordic Council of Ministers, and works to encourage and support Asian studies in the Nordic countries. In so doing, NIAS has been publishing books since 1969, with more than two hundred titles produced in the past few years.

UNIVERSITY OF COPENHAGEN

Nordic Council of Ministers

Printed by Printforce, United Kingdom